T0142812

Community and Climate Resilience
in the Semi-Arid Tropics

S. P. Wani • K. V. Raju

Editors

Community and Climate Resilience in the Semi-Arid Tropics

A Journey of Innovation

 Springer

Editors
S. P. Wani
Former Director, Research Program Asia
and ICRISAT Development Centre
International Crops Research Institute for
the Semi-Arid Tropics (ICRISAT)
Hyderabad, Telangana, India

K. V. Raju
Former Theme Leader, Policy and Impact,
Research Program-Asia
International Crops Research Institute for
the Semi-Arid Tropics (ICRISAT)
Hyderabad, Telangana, India

ISBN 978-3-030-29920-0   ISBN 978-3-030-29918-7   (eBook)
https://doi.org/10.1007/978-3-030-29918-7

This Springer imprint is published by the registered company Springer Nature Switzerland AG.
The registered company address is: Gewerbestrasse 11, 6330 Cham, Switzerland

Foreword

As we progress through the twenty-first century, the great challenge for humanity, of maintaining food and nutritional security, grows along with the Earth's population, the pressure on natural resources and climate change. This is particularly the case in Asia and Africa. Declining per capita the availability of water and land resources is threatening our ability to feed a growing human population, which is expected to reach over 9 billion by 2025. In India, the per capita water availability in 2011 has decreased to 1,545 cubic metres against the international threshold for water stress of 1,700 cubic metres. The National Institute of Hydrology estimates India's utilisable per capita water availability at just 938 cubic metres in 2010 and expects this to drop to 814 cubic metres by 2025.

Rainfed agriculture occupies 80% of the global arable land and contributes half the global food basket. While climate variability, resulting in droughts and floods, is a major driver of food insecurity in Asia and Africa, rainfed agriculture must continue to adapt in managing the inherent risks in food systems. The International Crops Research Institute for the Semi-Arid Tropics (ICRISAT) and our partners have found, through meta-analysis of watershed programmes in India, that rainfed agriculture in India is quietly revolutionising and that huge scope exists to enhance further the impacts of the watershed programmes – only 32% watershed performed above average.

On-station research at the ICRISAT has demonstrated over many years that the productivity of rainfed agriculture can be enhanced three- to fivefold over current yields through an integrated watershed management (IWM) approach. However, scaling-up adoption of IWM practices had been negligible despite the widespread on-farm demonstrations conducted in the States of Madhya Pradesh, Maharashtra, Karnataka and Andhra Pradesh.

An adoption survey undertaken in 1997 by a multidisciplinary team of scientists at the ICRISAT demonstrated the real potential of IWM approaches on major Indian soils (Vertic Inceptisols) covering 60 million ha. Subsequently in 1999, a pilot study was developed and implemented in Kothapally village (Adarsha Watershed), Telangana State, to demonstrate an innovative model of partnership. With the Kothapally community, the ICRISAT partnered with the state government, non-government organisations (NGOs), national research institutions such as the National Remote Sensing Centre (NRSC) and the Indian Council of Agricultural Research-Central Research Institute for dryland Agriculture (ICAR-CRIDA) and private sector companies to plan the implementation and monitoring of various watershed interventions.

A critical principle of the Kothapally experience was that beneficiaries paid in cash or in kind for the interventions that they received directly. The active participation of women and youth in watershed development and income-generating activities was essential. The two-decade experience in Adarsha Watershed at Kothapally (1999–2018) has resulted in accumulated lessons to guide India in its policies of watershed development and management at the national level.

This book, entitled *Community and Climate Resilience in Semi-arid Tropics*, is a substantial contribution by an ICRISAT-led consortium in the area of integrated watershed management that benefits smallholder communities in India. While it reports on benefits to millions of farmers in India, the flow-on impacts can already be seen in China, Thailand and Vietnam. This impressive contribution articulates scientific and policy measures for scaling-up appropriate community-based institutions and market linkages through public-private partnerships. The journey of Adarsha Watershed, Kothapally, serves as a lighthouse for guiding the development of rainfed areas in Asia and Africa.

I personally applaud Dr. S. P. Wani and Dr. K. V. Raju – who are not only the book's editors but also key leaders and implementors in the Kothapally story – for their meticulous efforts in bringing this book to publication. The same commendation goes to the chapter authors, most of whom worked in the fields with the Kothapally farmers and their community over the past decade. I am sure that this book will serve as a very valuable resource for development agencies, policy-makers, development investors, students and researchers.

I am particularly proud to see this publication from the ICRISAT and partners that documents how to enact ICRISAT's message of 'from science of discovery to science of delivery'. This publication reports good science, great impacts and, critically, their connections and lessons to improve our own practices in research. Well done to all the contributors, including our farmer and community partners in Kothapally.

Director General, ICRISAT   Peter Carberry
Hyderabad, India

Acknowledgements

We are very much thankful to all the stakeholders and consortium partners whose dedicated efforts made the Kothapally watershed as *Adarsha*, meaning a model watershed which served as lamppost for the development of watershed approach in India and parts of Asia, and to the tireless efforts of the dedicated team of scientists who initiated watershed research at the ICRISAT in 1976 and put up a strong foundation for watershed research in the institute and in India and the multidisciplinary team of scientist, namely Drs. P. K. Joshi, G. Algarsamy, T. J. Rego, Piara Singh and P. Pathak along with Editor S. P. Wani who initiated a multidisciplinary approach for watershed development at ICRISAT Center, Patancheru, based on the learnings from the adoption study undertaken in on-farm watersheds in Madhya Pradesh and Maharashtra. We sincerely acknowledge them as the team's efforts have showcased the success of integrated approach to the donors and policy-makers.

The unstinted support and efforts of Mrs. Rani Kumudini, IAS, Collector of Ranga Reddy District, is gratefully acknowledged as without her confidence and support, the journey of innovation in Kothapally would not have started as such. The financial support for this initiative from the Government of India through Drought Prone Area Programme (DPAP) through Government of Andhra Pradesh and also from the Asian Development Bank (ADB), Manila, Philippines (RETA 5812), enabled the team to establish the innovative pilot watershed and also piloted in other benchmark locations in India, Thailand, Vietnam and China. The support from the directors, DPAP, Ranga Reddy District, Government of Andhra Pradesh, since 1999 is acknowledged, especially the unstinted support by Dr. T. K. Sreedevi, Director, DPAP, Ranga Reddy District, who initiated the scaling up of the pilot in surrounding watersheds.

The consortium partners, namely Dr. H. P. Singh, Director, Central Research Institute for Dryland Agriculture (CRIDA), Hyderabad; Dr. R. Navalgund, Director, National Remote Sensing Agency (NRSA), Hyderabad; Dr. Shantha Sinha, CEO, MV Foundation, Hyderabad; and their team of scientists, contributed a lot in the development of *Adarsha* Watershed Kothapally. The farmers and the Watershed Association, the Watershed Committee and the self-help groups (SHGs) continually reposed their faith and provided all the support for evaluating new approaches and technologies as suggested by the scientists. The ICRISAT staff, notably Raghvendra Sudi, L. Jangwad, M. Babu Rao, D. S. Prasad Rao, Ch. Srinivas Rao, G. Pardhasardhy, K. N. V Satyanarayana, N. Sri Lakshmi, Y. Prabhakar Rao, Suchita Vithalani, M. Irsha and, late K. Srinivas, worked tirelessly to collect data, document results, prepare literature, organise visits and conduct training courses during the journey of innovation. During the process of innovation, several scientists, namely Drs. Joshi, P. K., Rego, T. J., Piara Singh, Pathak, P., Kesava Rao A. V. R., worked as a team initially, and later, other scientists who have contributed chapters worked and documented the learnings; their support and help are gratefully acknowledged. The unstinted support by Director General Shawki Barghouti and later Dr. William D. Dar who took the challenge to take the watershed research on farmers' fields was very critical and is thankfully acknowledged. During the journey of Kothapally Watershed, a large number of scientists and staff from the ICRISAT have contributed directly and indirectly, and their contributions are acknowledged. Last but not the least, the community in Kothapally always helped and ensured proper conduct of trials, without which, this book compilation would not be possible. We are highly thankful to all the chapter writers and other staff who have put together the learnings. We also thank Dr. Peter Carberry, Director General, ICRISAT, for writing the Foreword of this book.

S. P. Wani and K. V. Raju

Contents

Chapter 1
Need for Community Empowerment and Climate Resilience in the Semi-arid Tropics

S. P. Wani and K. V. Raju

Abstract The vast semi-arid tropics (SAT) area covering 120 million ha in Asia is also the home for 852 million poor and 644 million food and nutrition insecure people. Growing water scarcity and increasing land degradation in the dryland SAT areas are further aggravated due to impacts of climate change. In order to transform the dryland areas, innovative integrated watershed management model was developed and piloted by the International Crops Research Institute for the Semi-Arid Tropics (ICRISAT) in partnership through consortium approach, convergence with the government programs, collective action, and cooperation (4Cs) approach. How resilience of the communities was built through integrated watershed approach encompassing the livelihoods is described fully. The outlines of different chapters indicate briefly the strategy and various aspects including the process adopted and its impacts are covered.

Keywords Climate resilience · Integrated management · Watershed development · Drylad agriculture

1.1   Introduction

The semi-arid tropics (SAT), which covers 120 million ha area in Asia largely, is the home for the 852 million of poor people and 644 million nutritionally insecure people in Asia. Although, the SAT is blessed with weather where three crops can be grown, however, as water is the most scare resource in the region, large areas are cultivated by the farmers only with a single crop in a year. The impacts of climate change are also felt severe in this region largely because of increasing temperatures and growing water scarcity, which further get complicated with small land holders

S. P. Wani (✉)
Former Director, Research Program Asia and ICRISAT Development Centre, International Crops Research Institute for the Semi-Arid Tropics (ICRISAT), Hyderabad, Telangana, India

K. V. Raju
Former Theme Leader, Policy and Impact, Research Program-Asia, International Crops Research Institute for the Semi-Arid Tropics (ICRISAT), Hyderabad, Telangana, India

© Springer Nature Switzerland AG 2020
S. P. Wani, K. V. Raju (eds.), *Community and Climate Resilience in the Semi-Arid Tropics*, https://doi.org/10.1007/978-3-030-29918-7_1

and resource-poor farmers, who have neither access to the technologies to adapt to the impacts of climate change nor the financial resources to cope with. Under such circumstances, there is an urgent need to develop a model for adapting to the impacts of climate change and cope with the growing water scarcity, land degradation, and food production for sustainable development. To address the issue of improving the livelihoods of dryland farmers in the SAT, a model was planned and initiated in 1995 to harness the potential of dryland agriculture to bridge the yield gaps between the current farmers' yields and achievable potential.

The model, which was a holistic systems approach for enhancing crop productivity, was initiated on the ICRISAT campus in 1995. Based on the results of integrated watershed approach through multidisciplinary research by bridging the yield gaps, we demonstrated the potential to grow two crops successfully on large plots of Vertic Inceptisols. The approach was scaled up further in a 500 ha watershed in erstwhile Ranga Reddy district of Andhra Pradesh (Kothapally) (currently it is in Sangareddy District of Telangana state after the bifurcation of the state in 2014).

1.2 Focus of the Study

The focus of this study was on developing integrated holistic approach for harnessing the potential of rain-fed agriculture. In this approach, rainwater management through harvesting and recharging the groundwater was used as an entry point activity for increasing the productivity for the farmers through enhanced water use efficiency. To provide holistic and integrated solutions, the approach of consortium through building partnerships with different stakeholders like different research institutions (state, national, and international), development departments like Department of Agriculture, Department of Animal Husbandry, non-government organizations (NGOs), and Farmers' Organizations Community-based Organizations (CBOs), along with market linkages through private companies was adopted.

The focus of this initiative was on the 4Cs, namely, consortium, as explained above; convergence of various activities and schemes operated in the area; collective action of the farmers; and most importantly, the capacity building of the stakeholders mainly for adopting integrated approach in place of compartmental approach for providing solutions to the farmers. This particular approach of the 4Cs was expected to benefit the stakeholders through enhanced efficiency, environment protection, economic gain, and addressing the issues of equity (4Es) as the power of these 4Cs was far larger than the financial capital power. With this focus in mind, bridging the yield gaps for increasing the production and improving the livelihoods of the farmers through minimum environment damage for sustainable development was promoted through enhanced natural resource use efficiency. The success of this initiative was largely because of providing holistic solutions in a timely manner to the farmers and converging agriculture and allied sector activities for increasing the incomes of the farmers through capacity building; farmers got empowered and were wined away from the free inputs syndrome to ensure that the ownership is built amongst the farmers that will result into the demand-driven supply of knowledge/ technologies/inputs by researchers and development agencies rather than the supply-driven approach, which was not successful.

The main focus of this book is to document the learnings and share with other practitioners with an aim of scaling up in large areas to benefit millions of farmers in the country and other regions of the SAT in Asia and Africa. The success which we have recorded is not without trekking the difficult path dealing with communities who were accustomed to free dole outs and always were expecting something to get from the project as a passive partner in the initiative to a participatory approach for development, ensuring that they contribute in cash or kind to demonstrate/to take ownership and also ensuring that demand-driven-proven technologies are piloted to benefit the farmers. With this in focus for this book, the outline has been adopted as mentioned below.

1.3 Outline of the Study

For this innovative experiment of building the resilience of the community for climate change through innovative integrated watershed model, the chapters have been put in a simplistic manner for the reader to understand the whole process as well as the challenges and how the opportunities are harnessed resulting into impacts to benefit the farmers. Once the pilot was successful, it generated the demand from the surrounding villages because of the tangible economic benefits to the farmers, which clearly proved our first hypothesis that anything given free to the community does not get valued appropriately and in the process even the best of the technologies/products fail, and also the researchers/development workers cannot push the supply-driven technologies/products to the farmers as farmers are contributing and always look for value or satisfy themselves for getting tangible benefits from the technologies/products which are to be piloted.

- The first chapter deals with the need for community empowerment and climate resilience and the purpose of the study; provides the outline of the material practiced in a free flow for readers to understand; and describes in detail the methods used/adopted along with the impacts, the observations, and what ensured the success of the model.
- The second chapter deals with the farmers and ICRISAT's journey of innovation about how the Kothapally model was conceived based on the learnings of low adoption of on-farm watershed work, which was done through contractual participation of community and results from on-station multidisciplinary holistic experiments which enabled us to grow two crops without any supplemental irrigation on light black soils (Vertic Inceptisols), using sequential crops like soybean followed by chickpea and intercropping soybean with medium-duration pigeon pea using landform treatments for enhancing the harvesting of soil moisture storage and excess runoff water, which was used for recharging the

groundwater. The main focus of ICRISAT's journey is how the demand for a holistic and integrated approach emerged from the policy makers?.

- Based on this outcome and strategy, first and foremost, Chap. 3, titled "Climate Variability and Projected Change," explains the impacts of climate change and climate variability in the target eco-regions for which long-term weather data sets from the district were used and also presents the results. Once climate variability and its impact on the length of the growing period (LGP) was understood, the appropriate cropping systems were planned and piloted to address the issues of enhancing agricultural incomes in the Adarsha watershed, Kothapally.

- In addition to climate variability, soil health mapping was identified as an important constraint as farmers were not aware what they needed to apply for different crops based on the nutrient content in their soils. The results of soil health mapping in terms of physical, chemical, and biological properties was taken up and the results are presented in Chap. 4, along with providing an integrated soil management strategy. Soil-test-based nutrient management benefited the farmers through enhanced rainwater-use efficiency increasing the productivity per unit of rainfall, which really benefitted the farmers.

- Chapter 5 on rainwater management and eco-system services through integrated watershed management covers components of water balance and how these are affected due to integrated watershed management?. Integrated rainwater management interventions of in situ moisture conservation as well as ex situ rainwater harvesting for groundwater recharge as well as to be used for supplemental irrigation when rainwater is harvested within the field boundaries through integrated watershed development. This chapter also covers a number of ecosystem services provided through community participation, such as, provisioning, regulating, cultural/spiritual, and supporting.

- Chapter 6 deals with various cropping patterns/systems and crop intensification due to increased water availability in the watershed and also presents the results. Evaluation of improved crop cultivars as well as crop diversification using high-value crops with increased water availability to benefit farming families with enhanced incomes results are reported. The new cropping systems impacted changes in the cropping pattern and also increased net incomes for the farmers as well as sustainable use of natural resources. The results of these studies are presented in Chap. 6.

- In Chapter 7, the impacts of integrated watershed management are assessed using the economic surplus method. The results are reported as impacts covering social, economic, and biophysical effects that addressed natural resource issues for sustainable development, and social institutional impacts researched for the success of various initiatives resulting in tangible economic benefits to the community members are reported over the years from 1999 to 2016. The value chain for the agricultural products as well as allied sectors has been studied and proposed. This is the forward-looking approach as once the production and incomes have increased for the farmers, definitely they will have appetite for adopting value chain approach through collectivization, etc. The approach and possible potential value chains are discussed along with market linkages and strategies to minimize the post-harvest losses as discussed in Chap. 7.

- Chapter 8 deals with the use of digital technologies, including the implementation of satellite imageries since 1998, the land-use pattern , and the results of changes in the land-use pattern and the crop inventory. For geotagging the fields, a cell-phone-based app was developed and used successfully to map the farms along with the farmers' resource inventory, waterbodies, and land-use patterns during three seasons of the year.
- Chapter 9 deals with empowerment of women through income-generating micro enterprises, specifically through self-help groups (SHGs), in order to ensure their involvement in watershed activities. A number of income-generating initiatives, including the safe drinking water schemes and how over the years community evolved and took the ownership and initiative for new and improved lifestyles, are also reported. The role of empowered women in the sustenance of various watershed interventions is critical and a must for the success of sustainable management of integrated watershed approach.
- Chapter 10 deals with rural institutional governance mechanisms and how infrastructure (hardware as well as soft institutional mechanisms) has been developed and is being continued in the project area, although the project was withdrawn in 2003. The role of empowered rural institutions in the governance of watershed activities in this chapter provides the nuances of participatory and effective management of successful innovative watershed development model.
- The final chapter summarizes the whole concept of how the initiative was conceived based on a strategic research conducted on campus at ICRISAT and piloted in a village of 500 ha through community participation. Various interventions, the methods adopted, the institutional arrangements made, and the principle on which the project worked resulting into tangible economic benefits not only for the farmers but also for the team members, development workers, and development investors, which resulted in scaling up of this model from one village to thousands of villages in the country and also changed the watershed development guidelines at the national level.

Chapter 2
Adarsha Watershed, Kothapally, ICRISAT's Innovative Journey: Why, How and What?

S. P. Wani and K. V. Raju

Abstract The ICRISAT was working in watershed development since 1972 with Vertisol technology and piloted on farmers' fields in different agro-eco regions. However, it was not scaled up/adopted by the farmers in spite of the involvement of concerned state government agencies. In 1995, a multidisciplinary team of scientists' assessment of watershed studies in different agro-eco region pilot/benchmark sites indicated low adoption of Vertisol technology, although demonstrated on farmers' fields, was due to poor participation of the farmers as the approach was contractual participation and a one-size-fits-all approach was adopted. The new multidisciplinary experiment on station in Vertic Inceptisols demonstrated that using integrated watershed management approach these soils can be cropped during two seasons. Based on the demand of the district officials, Kothapally watershed was selected based on severe water scarcity, extent of rain-fed areas and the community's need and willingness to participate in the programme through full ownership/participation. The journey of innovation in Kothapally and how it became an exemplary (Adarsha) watershed with different strategies adopted are described. It evolved by the consortium of research institutions, government department, non-government organization and the farmers' community. The drivers of success are identified and the complete journey of innovation through a detailed timeline is covered in this chapter.

Keywords Holistic watershed · Innovation · Community empowerment · Watershed development · Climate change · Resilience · Drivers of success

S. P. Wani (✉)
Former Director, Research Program Asia and ICRISAT Development Centre, International Crops Research Institute for the Semi-Arid Tropics (ICRISAT), Hyderabad, Telangana, India

K. V. Raju
Former Theme Leader, Policy and Impact, Research Program-Asia, International Crops Research Institute for the Semi-Arid Tropics (ICRISAT), Hyderabad, Telangana, India

2.1 Background

The genesis of Adarsha watershed, Kothapally, can be traced back to the efforts of the team of scientists who realized that in spite of the long history of the watershed research by ICRISAT team since 1972 and also taking it to on-farm locations in different agro-climatic zones covering Andhra Pradesh, Karnataka, Maharashtra and Madhya Pradesh, the technology did not reach to the farmers in these states. If the technology has not benefitted the farmers in spite of strategic research on station and piloting in the on-farm sites, there was an urgent need felt to understand the reasons for the low adoption of such a technology which can double the farmers' incomes. To a certain extent why the study of Adarsha watershed, Kothapally, is covered in Chap. 1 in brief indicates the broad objective of this book. In this chapter, we dwell in detail on the genesis of the study; why it was undertaken; what were the compelling reasons to initiate this study and then how it evolved into a new strategic multidisciplinary study on the research station, piloting it to on-farm situation by changing the rules of the game of on-farm research?.

Further, what we did to take it to scaling up through adoption of the consortium approach to converge agriculture and related activities through collective action and capacity-building approach are reported. This chapter describes in detail the golden circle for integrated watershed approach of why, how and what.

2.2 Genesis of Adarsha Watershed, Kothapally, Why?

2.2.1 The Genesis of Adarsha Watershed

2.2.1.1 Rediscovering the Learning Cycle

The ICRISAT had undertaken watershed development approach since 1972 particularly for Vertisols (deep black cotton soils) which were left fallow during the rainy season, and farmers cultivated these soils on stored soil moisture during the post-rainy season (*rabi* season). Actual surveys of annual yields from farmers' fields in selected villages of peninsular India have been reported to be as follows:

Sorghum, (*Sorghum* bicolor)
Wheat (*Tritricum durum* Desf.)
Chickpea (*Cicer arietinum*)
Safflower (*Carthamus tinctorius* L.)
Chillies, dry (*Capsicum annuum* L.).

The reason for fallowing during the rainy season was as a risk mitigation strategy (Binswanger et al. 1980) to alleviate the waterlogging problem associated with Vertisols (Kanwar 1979; El-Swaify et al. 1985). The technology developed was called "Vertisol technology", which was a holistic farming systems approach, by following the watershed concept. The technology is comprised of several components, viz. contour field bunding; summer cultivation of soil taking advantage of off-season rains; broad bed and furrow (BBF) for addressing the issue of alleviating waterlogging as well as storing more rainwater as soil moisture (green water); dry seeding of seeds for most crops, except oil seed crops like groundnut (*Arachis hypogea*) and soybean (*Glycine max*) and small grains like millets (*Pennisetum glaucum*) and setaria (*Setaria italica*); balanced nutrient management; adoption of intercropping or sequential cropping to ensure double cropping (Krantz et al. 1976); rainwater harvesting; and integrated crop management, including pest management along with supplemental irrigation using harvested rainwater in the farm pond (Kampen 1982; El-Swaify et al. 1985). Long-term experiments conducted at ICRISAT centre, Patancheru, since 1976 clearly demonstrated that by adopting this technology using a number of crop combinations in intercropped as well as sequential crops, these soils can be cropped during rainy (*kharif*) and post-rainy (*rabi*) seasons even without any supplemental irrigation in assured rainfall regions. This path-breaking demonstration on farmers' field-scale (large) plots demonstrated that current farmers' crop yields were lower by four- to fivefolds as compared to the achievable crop yields under pure rain-fed situation. As per the farmers' practice (applying farm yard manure at 5 t/ha once in 2 years (Wani et al. 2003a)), cultivation of plots during the rainy season to keep plot weed-free and growing traditional *rabi* crops such as sorghum, safflower and chickpea on stored soil moisture yielded 1.1 t/ha as compared to 5.2 t/ha with improved management practice as mentioned above (Fig. 2.1, Tables 2.1, 2.2 and 2.3; Kampen 1982; Wani et al. 2001b, 2002b, 2003a). These levels reported from farmers' fields, sharply contrast with projected yields of up to 6 Mg/ha reported from research on several crops based on effective use of potentially available water (Kampen 1982; Swindale 1982). Not only the crop yields were higher by four- to five folds in improved management plots as compared to farmers' practice plot, but substantial improvement in soil physical, chemical and biological

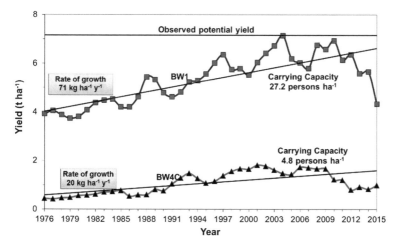

Fig. 2.1 Crop productivity of improved and traditional farmer's practice plots from long-term experiment at Heritage Watersheds at ICRISAT since 1976 (Source: ICRISAT 2017)

Table 2.1 Grain yields[a] (kg/ha) and gross monetary returns[b] (Rs/ha) for several crops at Kanzara village in 1979–1980

Watershed no.	Cropping system	Soil management	Sorghum	Pigeon pea	Cotton	Groundnut	Black gram	Gross value
Improved technology								
1	Sorghum/pigeon pea	Beds	2000	200	–	–	–	2630
1	Sorghum/pigeon pea	Flat	1470	210	–	–	–	2100
1	Cotton/sorghum/pigeon pea	Beds	760	60	630	–	–	3382
1	Cotton/sorghum/pigeon pea	Flat	560	60	490	–	–	2633
Existing technology[d]								
1	Cotton/sorghum/pigeon pea	Flat	20	20	180	–	–	767
1	Cotton/black gram	Flat	–	–	250	–	60	1112
Improved technology								
2	Sole groundnut	Beds	–	–	–	670	–	2177
2	Sorghum/pigeon pea	Beds	1470	200	–	–	–	2073

[a]The yields are based on small samples; actual threshing floor yields were somewhat (5–10%) lower

[b]The momentary values are based on the following prices at harvest time; sorghum Rs 105/100 kg, cotton Rs 385/100 kg, pigeon pea Rs 265/100 kg, groundnut Rs 325/100 kg, black gram Rs 250/100 kg

[c]"Improved" technology implies the use of recommended agricultural techniques in terms of seed, fertilizers, weed and insect control; existing; technology represents examples of the productivity attained with practices that are presently most common in the region

[d]Estimates of yields in adjacent fields (Source: Kampen 1982)

Table 2.2 Grain yields from a maize/pigeon pea intercrop system and a maize-chickpea sequential system compared with traditional rainy season fallow from deep Vertisol operation-scale watersheds at ICRISAT centre

Cropping system	Grain yields (mg/ha)				
	1976–1977	1977–1978	1978–1979	1980–1981	Mean
Maize/pigeon pea intercrop system					
Maize	3.29	2.81	2.14	2.92	2.79
Pigeon pea	0.78	1.32	1.17	0.97	1.06
Maize-chickpea sequential system					
Maize	3.12	3.34	2.15	4.18	3.20
Chickpea	0.65	1.13	1.34	0.79	0.98
Traditional fallow and single post-rainy season crop					
Chickpea	0.54	0.86	0.53	0.60	0.63
Sorghum	0.44	0.38	0.55	0.56	0.48

Source: El-Swaify et al. (1985)

Table 2.3 Physical properties of semi-arid tropical Vertisols under improved and conventional systems in a watershed at ICRISAT centre, Patancheru, India

Soil textural properties			
Texture	Improved system	Traditional system	SEM
Clay (%)	51	46	0.985
Silt (%)	22	22	0.896
Fine sand (%)	15	15	1.089
Coarse sand (%)	12	17	0.741
Gravel (%)	5	15	2.102
Hydrological properties			
Moisture retention (g g^{-1}) of 0–10 cm depth at 0.33 bar	0.35	0.33	
Moisture retention (g g^{-1}) of 0–10 cm depth at 15 bar	0.22	0.20	
Cum. infiltration in first 1 h (mm)	347	265	20.6
Sorptivity (mm h$^{-1/2}$)	121	88	14.6

Source: Pathak et al. (2011)

properties was also observed (Tables 2.4, 2.5a and 2.5b; Wani et al. 2003a). Similar results were observed in different studies during the same period. The success of watershed management largely depended on the community's participation. In a review (Joshi et al. 2000, 2008; Kerr et al. 2000) on the watershed projects in India, it was observed that most watershed projects could not address the issues of equity for benefits, participation of community scaling-up approaches, monitoring and evaluation measures. Moreover, most of these projects relied heavily on government investments. Also, most projects were structures driven (rainwater harvesting and soil conservation structures) and failed to address the issue of efficient use of conserved natural resources (soil and water) for translating them into increased systems productivity on large areas owned by smallholders mainly due to lack of technical support to such projects implemented by NGOs (Wani et al. 2001b).

Table 2.4 Biological and chemical properties of semi-arid tropical Vertisols under improved and conventional systems in a watershed at ICRISAT centre, Patancheru, India

Properties	System	Soil depth (cm)	
Soil respiration	Improved	723	342
(kg C ha⁻¹)	Conventional	260	98
Microbial biomass	Improved	2676	2137
(kg C ha⁻¹)	Conventional	1462	1088
Organic carbon	Improved	27.4	19.4
(t C ha⁻¹)	Conventional	21.4	18.1
Mineral N	Improved	28.2	10.3
(kg N ha⁻¹)	Conventional	15.4	26.0
Net N mineralization	Improved	−3.3	−6.3
	Conventional	32.6	15.4
Microbial biomass N	Improved	86.4	39.2
(kg N ha⁻¹)	Conventional	42.1	25.8
Non-microbial organic N	Improved	2569	1879
(kg N ha⁻¹)	Conventional	2218	1832
Total N	Improved	2684	1928
(kg N ha⁻¹)	Conventional	2276	1884

Source: Wani et al. (2003a)

Table 2.5a Grain yield under soybean/pigeon pea intercrop and soybean-chickpea sequential cropping system in a Vertic Inceptisol watershed at ICRISAT, 1995–1996 to 2003–2004

Soil depth	Mean grain yield (kg ha⁻¹)					
	Improved	Traditional	Improved	Traditional	Improved	Traditional
	Soybean		Pigeon pea		Soybean+pigeon pea	
Medium deep	1130	1150	920	940	2060	2080
Shallow	1060	1040	950	850	2010	1890
	Soybean		Chickpea		Soybean+chickpea	
Medium deep	1530	1450	1050	880	2570	2340
Shallow	1380	1350	640	560	2000	1930

Source: Singh et al. (1999)

Table 2.5b Grain yield under soybean/pigeon pea intercrop and maize-safflower sequential cropping system in a Vertic Inceptisol watershed at ICRISAT, 2004–2005 to 2011–2012

Soil depth	Mean grain yield (kg ha⁻¹)					
	Improved	Traditional	Improved	Traditional	Improved	Traditional
	Soybean		Pigeon pea		Soybean + pigeon pea	
Medium deep	1159	1080	918	832	2132	1962
Shallow	1028	856	701	590	1795	1501
	Maize		Safflower		Maize + Safflower	
Medium deep	4901	4623	864	682	5765	5305
Shallow	4301	3437	635	441	4936	3878

Source: Singh et al. (1999)

2.3   On-Farm Evaluation of Watershed Technologies (Vertisol Technology)

Following the excellent results observed in long-term on-station plots, scientists decided to take this technology package for on-farm evaluation in Andhra Pradesh, Karnataka, Madhya Pradesh and Maharashtra. Field-scale watersheds were selected and with the farmers contracts were made for undertaking the demonstration of Vertisol technology as a package comprising the components mentioned above as a holistic system. Although, the proposed approach was a holistic farming system approach, the implementation was not truly holistic. ICRISAT staff were posted at sites to collect all data as well as proper implementation of all the components of the Vertisol technology. During the demonstration phase the results were excellent as farmers could grow two crops and their family incomes increased more than two-folds (Walker et al. 1983) and also generated employment for longer period for the family members as well as hired labourers (Table 2.4). Once the technology was demonstrated for 4–5 years, scientists withdrew the technical support as well as ICRISAT staff who used to undertake implementation of various activities as planned. It was anticipated that the farmers on whose fields the technology has been demonstrated and also to others disseminated by conducting Field Days the technology adoption would increase, as economically the technology was excellent with more than 100% increase in incomes and government departments in the states were also associated with the demonstrations. There were sporadic reports about non-functioning of Vertisol technology package as such, and it was thought that the technology which is suitable for deep black cotton soils covering 12 million ha in India which are prone to waterlogging was applied by the farmers/officers/researchers to inappropriate adoption zone (shallow black soils with less rainfall, etc.), and that's why such reports were emerging. At the same time, the on-station demonstration plots were showing good successful results over a long period.

2.3.1   Revisit to On-Farm Watersheds to Understand Low Adoption of Technologies

In 1995, under a newly formed system project III dealing with medium rainfall zone, the multidisciplinary team of scientists (natural resource economist, soil physicist, land and water management scientist, agronomist and soil biology cum plant nutrition scientist) decided to assess the reasons for poor adoption of Vertisol technology. The team visited Raisen Watershed in Madhya Pradesh as well as Aadgaon Watershed in Maharashtra and interacted with the farmers who had participated in the on-farm demonstrations as well as scientists from the State Agriculture University and Water and Land Management Institute (WALMI) in Bhopal and Aurangabad. This was the first time that a multidisciplinary team of scientists from ICRISAT with local region scientists together interacted with the farmers 15 years after withdrawal of the project to understand the reasons for failure or low adoption of Vertisol technology in the regions where technology demonstrations were conducted.

The multidisciplinary team was unique as they were willing to learn afresh from the farmers the reasons for low adoption of technology. Their genuine urge to understand the reasons without any attachment with the technology helped to come out with the learnings based on the interaction with the farmers who undertook demonstrations as well as the surrounding farmers and the scientists working in the region. The results were eye opening for the team as lot of new learnings were emerging during the evening frank discussions amongst the team members from different perspectives. The purpose of the mission was not to find faults with the earlier thinking or implementation but a real urge to make the watershed technology (Vertisol technology) popular amongst the farmers to benefit them as evident from the strategic research in Heritage Watersheds at ICRISAT campus. The major findings indicated that even in the same regions which were selected for demonstrations, except improved seeds and fertilizers, other components of the technology were not even seen on any fields. Even field bunding, which was undertaken on contours, was demolished, and no rainwater harvesting in farm ponds, no summer cultivation, nor dry seeding was followed by any of the farmers. The team was surprised that no farmers were following the critical components of the technology except the improved seeds and fertilizers, which were more largely due to persuasion, and other incentives provided by the private companies. During the detailed discussions amongst the team members as well as documenting the process, it was observed that the approach adopted for conducting on-farm demonstrations was a contractual collaboration with the farmers as farmers were paid the charges for their land use, inputs were provided by the institute, all field operations were undertaken by the institute staff located on site and farmers were getting all the benefits of increased crop productivity, plus getting the attention and popularity in the village during the Field Days. This learning loop opened the eyes of the team and initiated the thinking how watersheds can be popularized and farmers could benefit from the technologies developed by the researchers?.

2.4 How Adarsha Watershed, Kothapally, Was Conceived?

2.4.1 Designing New Multidisciplinary Experiment for Technology Development for Vertic Inceptisols

After learning from the survey and looking at the long-term experiments conducted in the Heritage Watersheds, the team felt that the Vertisol technology application domain in India is only less than 12 million ha as many of Vertisols do not get waterlogged as they are in low rainfall zones or have a good drainage. However, in 60 million ha Vertic Inceptisols (shallow and medium deep black soils) in India, the institute has no technologies to demonstrate that two crops can be grown on these soils without supplemental irrigation. Equipped with the eye-opening revelations from the survey, availability of proven Vertisol technology and the need to develop technology to grow two crops on Vertic Inceptisols, the team started planning an experiment to demonstrate that even with 800 mm annual average rainfall on medium to shallow black soils, two crops can be grown with appropriate technologies and crop combinations. In addition to the learnings from the assessment survey and the need for developing suitable technology for unlocking the potential of Vertic Inceptisols, there was another important but compelling reason to join hands for a multidisciplinary experiment at ICRISAT centre. In the new organization, the NRM programme was leading three systems projects, and for assured medium rainfall zone PS III (production system III) project, the operational funds for the team of five scientists were very meagre with which it was not possible for the individual scientists to run independent research experiments.

With this background, the lead was taken to design a multidisciplinary experiment to develop technology for double cropping of shallow to medium depth black soils. The team started its search for a suitable site to be developed as a research scale watershed and zeroed on a field (BW 7) which had varying soil depth from 75 to 5–10 cm along the slope mimicking the real-world situation in the watersheds. The team designed the main treatment as soil depth (three depths, viz. deep, medium and shallow) and sub-treatment as landforms (two, viz. broad bed and furrows and flat on contour) and the sub-sub-treatment as cropping systems (two, viz. sequential soybean (later replaced with maize to avoid continuity of legumes) followed by chickpea and soybean intercropped with medium-duration pigeon pea). The team consulted a statistician to avoid the later complications to undertake analyses of data to test the designed hypotheses. Each scientist collected the needful data for their study from the same experiment. Once the team decided and finalized the design, it moved along to plant the first crop in 1995 in a newly started experiment in BW 7. As the team members were on board from the beginning, the team looked after the experiment regularly and many a times together to discuss and address the on-ground issues during the field visits. The main plots of soil depths were separated by contour bunding, and all the bunds were planted with *Gliricidia sepium* saplings to address the issue of low soil carbon content using the N-rich organic matter generated in the field. The crop residues were composted in the compost pits.

Automatic weather station near the field provided all daily weather data, and each main plot was equipped with automatic hydrological gauging station to monitor runoff and soil loss from the main plots. The excess rainwater was harvested in two farm ponds and all the waterways were fully grassed. All the operations were undertaken using the bullock-drawn *tropicultor*. The success was evident from the first season itself as the total system productivity was around 3–4 t per ha as against 0.5–0.8 t/ha on farmers' fields depending on soil depth without any supplemental irrigation (Singh et al. 1999, Tables 2.5a and 2.5b). Soon this became one of the best spots for the institute visitors to see the systems research with all the scientific data collected explaining various processes of rainwater management, runoff and groundwater recharge, crop growth parameters, productivity, integrated soil fertility management and soil biology and most importantly to manage green water (soil moisture) efficiently for enhancing crop productivity and profitability for the farmers while minimizing land degradation.

The team leader was interacting with the important visitors and also liaising with the government officials from the district (Ranga Reddy District in erstwhile Andhra Pradesh). During one of the visits, the officials from the Asian Development Bank, Manila, Philippines, visited ICRISAT and during the field visit, visited BW 7 integrated system's approach experiment during 1997–1998 when the experiment was already in third year. The excellent results demonstrating the technology to unlock the potential of rain-fed agriculture in the tropics attracted the attention of the ADB officials. During the field visit, the ADB officials enquired about scaling-up plans for the technology which is ripe to take to the farmers' fields. The team expressed their confidence but highlighted the scarcity of funds to take the technology to the on-farm testing. During the wrap-up meeting, the ADB officials indicated that the bank will be happy to support a scaling-up pilot for the BW 7 technology, which is matured enough in their opinion. The ADB asked the team and the institute to submit the proposal covering three countries with varying rainfall situations. This was the first sign of success of the multidisciplinary system's approach for the team which pushed their confidence to greater heights.

At the same time, the collector of Ranga Reddy district (Mrs. Rani Kumudini, IAS), with whom the team leader was liaising, requested the team's help to plan a watershed for the land to be allocated to 12–15 landless families in the district. The team decided to survey the available land and plan the watershed before distributing the *pattas* (land ownership papers) to the families. The team planned the common waterways, the contour bunds to divide the land into equal land parcels and a place for the farm pond. The mapped watershed plots were distributed by the government to the landless families, and a new way to manage rainfall in the common/wastelands in the state was introduced. Following this exercise then the collector requested to organize a training course for the watershed committees in the district at ICRISAT. During the inaugural session of the training course, the honourable minister of agriculture was the chief guest. After the inaugural session the minister visited the on-station watershed experiments along with the collector. During the lunch discussions, the honourable minister said, "you have excellent technologies on the station. You should demonstrate these technologies outside the compound of the institute. The government will be willing to provide the needed funding". The collector was told to take this initiative forward in the district and help the farmers with ICRISAT developed technologies.

Following these discussions, the collector asked ICRISAT team to select a 500 ha watershed in the district as per their choice and demonstrate the technologies on pilot scale. The government indicated that funding will be provided as per the needs. This was the second success for the team following the ADB's willingness to support the scaling-up initiative. The leadership deliberated the options and it was decided that if the scaling-up model has to be developed, then it would be better to work within the existing government system instead of taking the funding and developing a pilot which will again face the challenges of enabling institutional and policy with the government setup. As the ADB funding was on the horizon to undertake strategic research as well as cover the team's cost, a calculated risk was taken and indicated to the collector that normal funding for the watershed programme under the Drought Prone Area Programme (DPAP) needs to be provided for the pilot but with a caveat that being the pilot to be developed as a model, the new initiatives, approaches and implementation arrangements need to be permitted overriding the existing government guidelines which could be restrictive for new interventions. The district collector readily agreed to this approach and said "you will have all permissions to develop a model as you like and no questions will be asked by the officials for any deviations made to the existing policies". That's how a foundation for the new watershed model was laid by the ICRISAT and the district administration of the Ranga Reddy district under the leadership of the collector.

2.5   What We Did to Establish Adarsha Watershed, Kothapally?

Once we had these two offers for developing a model for the new watershed management approach from the undivided government of Andhra Pradesh and also from the ADB to demonstrate the integrated watershed management technology, the team moved ahead to plan and take up the challenge thrown at us by the honourable minister of agriculture of Andhra Pradesh.

2.5.1   Selection of Kothapally Watershed Based on the Learnings from the On-Farm Survey by the Multidisciplinary Team of Scientists

For selecting the watershed the team had followed a set of criteria such as:

maximum cultivable area in the village should be rain-fed and water scarcity should be the main concern of the villagers (demand driven) for developing agriculture;
poverty, which is directly associated with availability of water in rural India, should be there;
little area under irrigation using groundwater;
people should be willing to collaborate as per the terms;
good local leadership should be available;
the site should be accessible during the rainy season and a representative for the district/region in terms of soil type, rainfall, socioeconomic parameters and around ICRISAT campus so that visitors can be taken to the site as and when needed; etc.

Once the criteria were developed, a team of ICRISAT scientists along with the representative from the DPAP for the government of Andhra Pradesh visited a set of three villages around the ICRISAT campus. The three villages, viz. Kothapally, Parveda and Urella, in the Ranga Reddy district were evaluated based on the criteria. In each village after the transect walk with the villagers, a meeting was held with the villagers. The purpose of the visit of the team was elaborated, details of the project were discussed and people's feedback/reactions were noted. Based on the cumulative score of the team members, Kothapally was ranked as the first choice for developing a model, Urella was the second and Parveda village was the last one which, was a predominantly cotton-growing village with groundwater availability.

Once the first and second choices of villages were identified, again a second detailed consultation with all the villagers by conducting a village meeting was undertaken. During this meeting it was highlighted how their village has been selected as potential village for the project. However, the criteria of people's willingness to collaborate were retested by detailing the terms and conditions of the collaboration. It was made clear to the villagers that:

- In this project except knowledge and technical support by the team of scientists no other inputs will be provided by the project free of cost. Each participant in the project will have to contribute their share in cash or kind (by those farmers who cannot contribute upfront cash). *This was the first new parameter included in the project.*
- The whole village should be united as one as far as their project activities are concerned and political association with particular political party should not interfere in the project.
- The villagers will need to select unanimously the watershed committee (WC) members as per the criterion provided by the DPAP department officials within 2 weeks.
- The WC will have to be registered with the Department of Cooperatives, GoAP, and bank account has to be opened by the WC in the nearest bank.
- All payments for the watershed activities undertaken will be through bank cheque payments, and transparency will have to be maintained for all the expenses from the project as well as the contributions made by the members.
- Most importantly, whenever the team is visiting and a specified time is indicated, community members should be present on time as during the second meeting, in spite of fixing the time, people were to be called and gathered after the team arrived, which should not be the case in future.
- In future, no ICRISAT team member will accept tea, snacks, lunch or any favours from the villagers (*this was the second new parameter included in the process*) to avoid any misconception about favouritism shown by the project team for specific activities for the influential people in the village.

Once the agreement was reached on the modalities of collaboration, then Kothapally was finalized as the final site for the new model of integrated watershed development. This process was to ensure that the community members were proactively engaged from the inception phase of the project to avoid the mistake of contractual participatory research undertaken during the earlier phase of on-farm watershed development and ensure that the participation is at the highest order of collaborative participation as against the contractual, consultative or cooperative participation of the community. Once the community agreed to follow the project guidelines, the leader had promised the community in 1999 that if the community implement the proposed activities fully and wholeheartedly, we assure that the name of Kothapally will be not only across the country, but it will be on the world map for the new model of integrated watershed development.

2.5.2   Formation of the Consortium for Implementing Integrated Watershed Development Model

In the meantime, parallel to the selection process, we were also identifying suitable and appropriate partners to support the implementation of the model watershed. We zeroed on the best non-governmental organization (NGO) working in the area. We came across a good reputed NGO, MV Foundation, that was doing excellent work in the area of child labour eradication and ensuring that all schoolgoing-age children are in school and not working in homes or fields. We met the chief founder of MV Foundation, Mrs. Shantha Sinha, who was a well-known social worker internationally and briefed her about the proposed model to be developed. She was excited about the project but said that MV Foundation has no expertise in agriculture or watershed development. We indicated that we would like to harness their skill for social mobilization as well as maintaining transparency in the financial matters. She agreed to become a partner, but when we indicated about the association of DPAP, she said she is not keen to join hands with the government machinery as NGOs in general are looked suspiciously by the government officers for financial matters and she would not like to deal with financial management.

We called a meeting of DPAP and NGO officials together at ICRISAT and indicated that the consortium will work on harnessing the strengths of the partners and other aspects can be resolved with the guidance of the madam collector. After the meeting we discussed with madam collector the concerns raised by the NGO officials and it was agreed as an exception that the money will be passed through the WC and DPAP officials will assist as a partner in recording measurement book (MB) and calculate the payments for the works completed at regular interval. Through this arrangement the problem of handling government money by the NGO was surpassed, and the DPAP became active partner to work with the community, which was a new mode of working in the watershed programmes with clear role for each partner. *This was the third parameter included* in the project implementation.

2.5.2.1   Approach Adopted

The new approach comprised several new components of the farmers' participatory consortium approach for community watersheds as follows:

- Involvement of government authorities in the consortium from the beginning
- Formation of consortium of local, regional, national and international research and development institutions for providing technical support to the NGOs and farmers

- Refinement of technologies and on-farm strategic research experimentation by farmers with technical support from the consortium partners and farmers' contribution in cash or kind

The process/approach is depicted in Fig. 2.2.

2.5.2.2   Integrated Watershed Management–ICRISAT's Innovative Consortium Model

A new farmer participatory holistic consortium model for efficient management of natural resources for improving rural livelihoods emerged from the lessons learnt through long-term watershed-based research led by International Crops Research Institute for the Semi-Arid Tropics (ICRISAT) and national partners. The important lessons learnt from earlier watershed-based research were:

- Generally, researchers worked with progressive farmers, and as a result equity for benefits to smallholders and landless was compromised.
- Contractual mode of participation resulted in low and passive community participation and sustainability for managing the watersheds after the project was lacking.
- Community participation was low in watersheds as a small proportion (5–10%) of farmers having access to groundwater were only deriving benefits from the interventions.
- Emphasis was on establishing/demonstrating pilots and not scaling up as it was supposed to happen automatically with dissemination process (trickle-down effect).
- Evaluation of the watershed interventions was undertaken as a postmortem activity and not as a concurrent learning process.
- Scientists were working independently for pilots and as a result technical support for most development projects implemented by NGOs/government departments was lacking to address the issues holistically.

The important components of the new model, which were distinctly different from earlier models (Wani et al. 2002a), were:

1. Demand-driven approach was adopted for selecting the watershed and the farmers collectively identified and prioritized the problems for possible technical interventions.
2. Consortium approach involving needed research (national, international and local) and development institutions along with government departments was adopted from the beginning.
3. Participatory planning and implementation of watershed research and development with the involvement of all stakeholders. Farmers' groups selected the sites for rainwater harvesting structures, cropping systems and varieties or technologies with technical support from the consortium partners.
4. New science and technology tools such as remote sensing (RS), geographical information system (GIS), digital terrain modelling (DTM), soil health map-

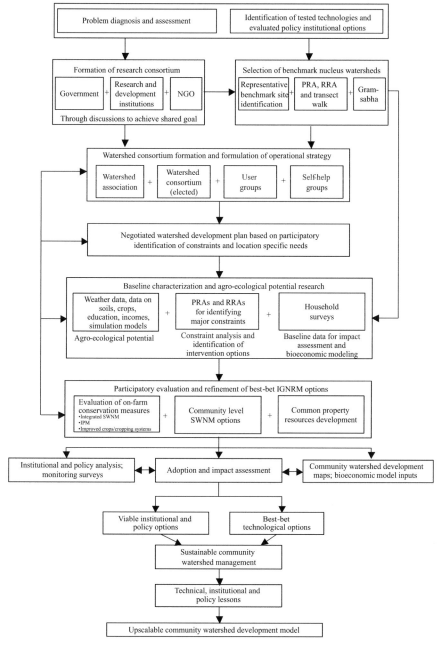

Fig. 2.2 Process of participatory consortium approach through community watersheds. (Source: Wani et al. 2002b and 2003b)

ping including soil depth, integrated nutrient management and crop simulation models were adopted along with new knowledge dissemination methods.

5. Linking successful on-station watersheds and on-farm watersheds for strategic research enabled the farmers as well as researchers to think differently to solve their problems.

6. In place of mere soil and water conservation, a holistic system approach for watershed management programme for livelihood improvement to benefit all the community members who were deprived of the project benefits in earlier programmes was introduced.

7. Increased individuals' participation by ensuring tangible private economic benefits to small farm holders, landless families, youths and women through income-generating activities. The emphasis is on in situ conservation of rainwater and translating benefits of increased soil water availability through integrated genetic and natural resource management (IGNRM) approach.

8. The islanding approach through establishment of a micro-watershed within the watershed which served as a site of learning within the village itself and also to build the confidence of farmers by undertaking research.

9. For technical development and inputs on individual/private land users pay 50% (with incentive) and for community-based interventions largely government pays with 10–30% contributions from beneficiaries.

10. For scaling-up and technology dissemination use of benchmark sites as training sites for partners and farmers and for sensitizing the policymakers with an intention to develop scaling-up model for the successful pilot.

11. Cost-effective and environment-friendly soil, water, nutrient, crop and pest management for wider and quick adoption to ensure tangible economic benefits to the community.

12. Validation of conventional/traditional knowledge of the community for amalgamation with new scientific knowledge for sustainable management of natural resources.

13. Collective action of the community along with capacity building of local farmers and NGOs for impact-oriented interventions and for dissemination of technologies. Strengthening of community institutions through dedicated efforts for ensuring sustainability of the interventions made.

14. Continuous monitoring, participatory evaluation and learning by researchers and stakeholders to assess the overall performance of watershed management for midcourse corrections.

2.5.3   Adarsha Watershed, Kothapally

The watershed selected is located in Kothapally village (longitude 78° 5′–78° 8′E and latitude 17° 20′–17° 24′ N) in Ranga Reddy district, Andhra Pradesh (erstwhile undivided), nearly 40 km away from ICRISAT centre, Patancheru. The team and the villagers decided to name it as Adarsha (an exemplary) watershed, indicating from

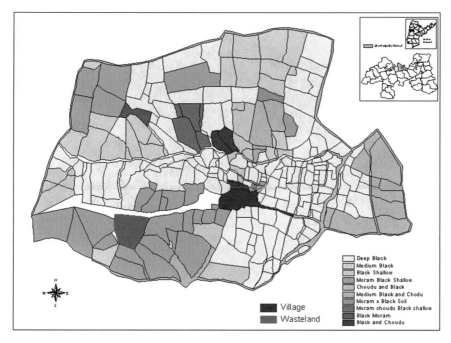

Fig. 2.3 Soil types of Adarsha watershed, Kothapally. (Source: Shiferaw et al. 2002)

the beginning itself that everything done in this watershed has to be the best possible. It is spread over 465 ha of which 430 ha is cultivated and the rest is wasteland. The watershed is characterized by undulating topography with an average slope of about 2.5%. Soils are predominantly Vertisols and associated soils (90%) (Fig. 2.3). The soil depth ranges from 30 to 90 cm with medium-to-low water-holding capacity (Fig. 2.4). The total population in Adarsha watershed was 1492 belonging to about 270 cultivating and 4 non-cultivating families in 1998. The average landholding per household was 1.4 ha (Shiferaw et al. 2002).

2.5.3.1   Consortium Approach

An innovative consortium model with partnership of institutions for technical backstopping, as against implementation by a single institution, was adopted. The ICRISAT, M. Venkatarangaiya Foundation (MVF), an NGO, Central Research Institute for Dryland Agriculture (CRIDA), National Remote Sensing Agency (NRSA now referred to as National Remote Sensing Centre), District Water Management Agency (DWMA) and Ranga Reddy district of Andhra Pradesh Government along with farmers formed the consortium (Fig. 2.5) for project implementation (Wani et al. 2001a). All the partners were working individually or in partnership with another institution to conserve rainwater and manage the watershed sustainably with clarity of role for each partner along with the responsibility to deliver.

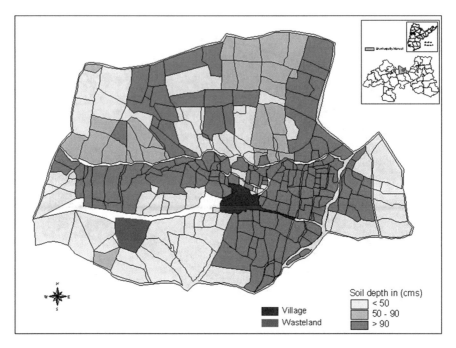

Fig. 2.4 Soil depth map of Adarsha watershed, Kothapally. (Source: Shiferaw et al. 2002)

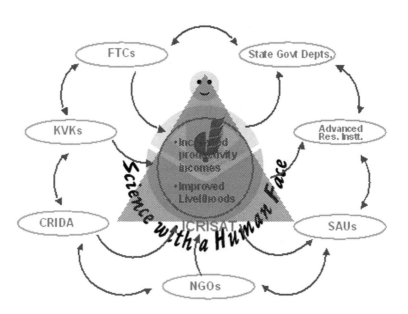

Fig. 2.5 Farmer participatory consortium approach for integrated watershed development. (Source: Wani et al. 2001b)

2.5.3.2 Promoting Community Participation

The participation of local community, i.e. farmers, is a must for successful impact of watershed management (Samra 1997; Joshi et al. 2000, 2006; Kerr et al. 2000). A successful partnership based on strong commitment with state and local agencies, community leaders and people is desirable. To promote community participation in the watershed for site selection, implementation and assessment of activities, various committees/groups were formed. It was recognized that to shift the community participation from contractual to consultative and collegiate mode, tangible private economic benefits to individuals are a must. Such tangible benefits to individuals could come from in situ rainwater conservation and translated through increased farm productivity by adopting integrated genetic and natural resource management (IGNRM) approach. Most importantly, from the beginning stage, watershed selection to technology selection; crops, systems and varieties selection and evaluation; monitoring and evaluation of watershed activities were done in full participatory mode. No full subsidies for investments on individual's farms for technologies, inputs and conservation measures were provided by adopting the principle that "users pay". Once the individuals could realize the benefits of soil and water conservation, they come forward to participate in community activities in the watershed through various organized groups as follows:

Watershed association: All the 274 farmers (landowners and landless families) were members of the watershed association. The association was registered under Registration of Societies Act and is a sovereign body that decides every activity of the watershed.

Watershed committee: It was the executive body of the association and was headed by a chairperson who was unanimously elected. A secretary, who maintained the records, and eight members representing different sections of the community were other members of the watershed committee.

Self-help groups: Self-help groups were formed to undertake watershed management activities.

User groups: User groups were formed to manage (operate and maintain) water harvesting structures.

Women self-help groups: Women were empowered to form self-help groups to undertake village-level enterprises for income generation. Ten such groups with 15 members each took up vermicomposting as an enterprise in Kothapally village initially. Later, other income-generating activities such as nursery raising, livestock rearing, spent malt as animal feed distribution, kitchen gardening with vegetables and small businesses in the village were undertaken.

Further developments in Adarsha watershed, Kothapally, are covered in the subsequent chapters in detail.

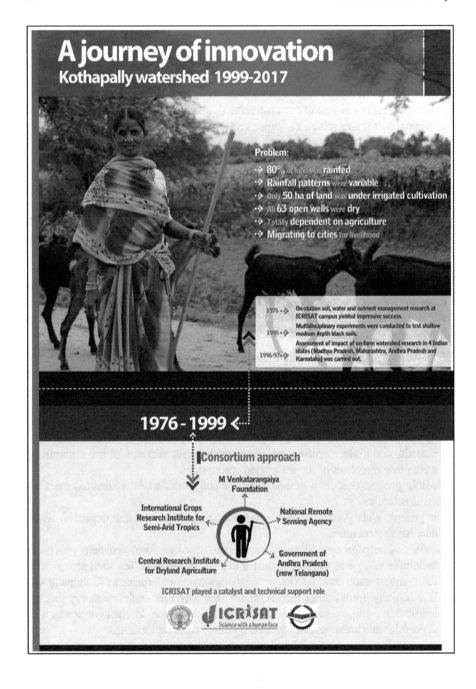

The **Broad-bed and Furrow** method controlled soil erosion. Tropicultor was used for planting, fertilizer application and field bunding.

Wilt-tolerant pigeonpea cultivar ICP 87119 (Asha) was introduced as an entry point activity for participatory trials.
····▷ 1999

Runoff Gauging Station was set up at Kothapally to monitor runoff and soil loss from treated and untreated catchment.
····▷ 1999

Automatic weather station was set up at Kothapally village on a farmer's field to collect data on soil, temperature, rainfall and wind speed.
····▷ 1999

Farmers started to **diversify**, grow flowers and vegetables in small areas of land with availability of water.
····▷ 1999-2000

Community members mobilized to form groups built and maintained **low-cost water harvesting structures** like earthen check dams, percolation pits, gully plugs, sunken pits and well recharge pits.
····▷1999-2000

Four surrounding villages of Kothapally requested similar support from ICRISAT – first scale-up on demand.
····▷ 2000

Soil and water ⁞ **Interventions**

····▷1999 - 2000◁····

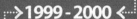

Impacts

·▷ Nearly **53,475m³** of water storage capacity (5-8 times the storage capacity).

·▷ Nearly **₹2 million** in construction costs and **₹20,000** in annual maintenance was saved.

·▷ All **37 recharge wells** have become functional through runoff water; a total of **200 ha** were irrigated in *rabi* (postrainy season) and **100 ha** in summer.

·▷ Average runoff in treated catchment reduced by **13%** compared to untreated catchment; soil loss reduced by **45%**.

·▷ Increased **water** availability round the year.

Gliricidia plants on the **field borders** provided nitrogen-rich manure and prevented soil erosion.
⋯⋯▷1999-2001

Nutrient budgeting corrected Nitrogen, Phosphorus and Potassiumdeficiencies, **detailed soil characterization** helped make Boron and Sulphur amendments to the soil.
⋯⋯▷1999-2001

Pest control measures like village level Helicoverpa Nuclear Polyhedrosis Virus production, installing bird perches, shaking pigeonpea plants to control pod borer and *rhizobium* inoculation increased crop productivity.
⋯⋯▷1999-2001

With **change in cropping patterns** and **intercropping**, farmers were able to grow 2-3 crops per year in a small patch of land (where they were earlier growing vegetables and flowers).
⋯⋯▷1999-2001

Vermicomposting turned into an **income generating activity** for women; crop yields increased, and use of pesticides fell.
⋯⋯▷1999-2001

Productivity enhancement ⋮ Interventions

⋯▷1999 - 2001 ◁⋯

Impacts

·▷ Average crop yield of maize alone increased by 2.2 to 2.5 times.

·▷ Pigeonpea production increased to 900 kg/ha as against 200 kg/ha in 1998.

·▷ Farmer's average income increased by 3 folds in 2010 compared to ₹ 25,000 in 1998. Even during the drought year (2002), increase in income was 1.5 folds and people from surrounding villages were coming to Kothapally in search of work.

·▷ Integrated pest and nutrient management practices + drip irrigation helped improve crop productivity; net income ranged between ₹ 25,000 and ₹ 72,000 per acre.

·▷ Kothapally villagers now own 35 autos, 2 transport vans, 4 lorries and 9 tractors.

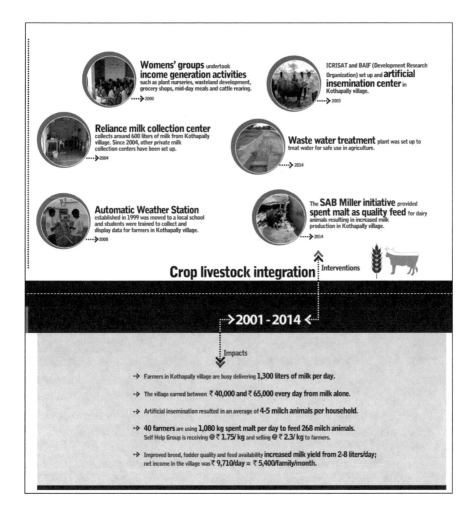

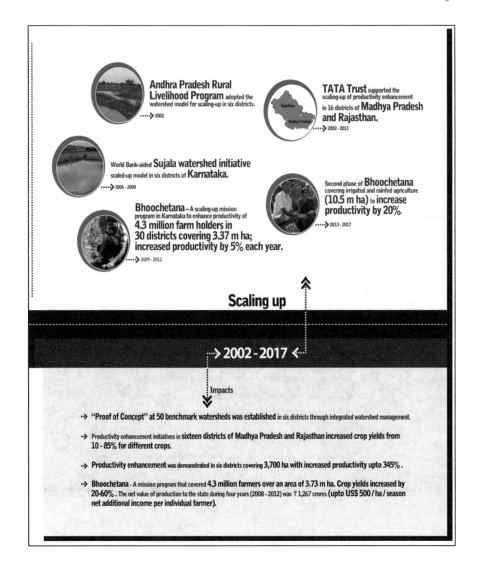

References

Binswanger, H. P. Virmani, S. M., & Kampen, J.. (1980). *Farming systems components for selected areas in India: Evidence from ICRISAT*. Res. Bull 2. International Crops Research Institute for the Semi-Arid Tropics (I(.RIS,I r). Patancheru, A.P., India.

El-Swaify, S. A., Pathak, P., Rego, T. J., & Singh, S. (1985). Soil management for optimized productivity under rainfed conditions in the semi-arid tropics. *Advances in Soil Science, 1*, 1–64.

ICRISAT (International Crops Research Institute for the Semi-Arid Tropics). (2017). Small agricultural watersheds at ICRISAT. Flyer.

Joshi, P. K., Tewari, L., Jha, A. K., & Shiyani, R. L. (2000). Meta-analysis to assess impact of watershed. In *Workshop on institutions for greater impact of technologies*. New Delhi: National Centre for Agricultural Economics and Policy Research.

Joshi, P. K., Pangare, V., Shiferaw, B., Wani, S. P., Bouma, J., & Scott, C. (2006). Socioeconomic and policy research on watershed management in India: Synthesis of past experiences and needs for future research. *Journal of SAT Agricultural Research, 2*(1), 1–81. ISSN 0973-3094.

Joshi, P. K., Jha, A. K., Wani, S.P., Sreedevi, T. K., & Shaheen, F. A. (2008). Impact of watershed program and conditions for success: A meta-analysis approach. Global theme on agroeco-systems report no. 46. *Monograph*. International Crops Research Institute for the Semi-Arid Tropics, Patancheru, Andhra Pradesh, India.

Kampen, J. (1982). *An approach to improved productivity on deep Vertisols.. Information Bulletin No. 11*. Patancheru: International Crops Research Institute for the Semi-Arid Tropics.

Kanwar J. S. (1979). *Research at ICRISAT-an overview*. Proceedings of Inaugural Symposium on tropics prospects and retrospect. Proceedings of international symposium on Development and Transfer of Technology. ICRISAT. Patancheru P.O., AP, India.

Kerr, J., Pangare, G., Pangare, V. L., & George, P. J. (2000). *An evaluation of dryland watershed development in India.. EPTD discussion paper 68*. Washington, DC: International Food Policy Research Institute.

Krantz, B. A., Virmani, S. M., Singh, S., & Rao, M. R. (1976). Intercropping for increased and more stable agricultural production in the semi-arid tropics. In *Proceedings, Symposium on Intercropping in the Semi-Arid Areas* (pp. 15–16). International Development Research Centre (Canada), 10–12 May 1976, Morogoro, Tanzania.

Pathak, P., Wani, S. P., & Sudi, R. R. (2011). Long-term effects of management systems on crop yield and soil physical properties of semi-arid tropics of Vertisols 2(4), pp. 435–442 openly accessible at http://www.scirp.org/journal/AS/. https://doi.org/10.4236/as.2011.24056.

Samra, J. S. (1997). *Status of research on watershed management*. Dehradun: Central Soil and water Conservation Research and Training Institute.

Shiferaw, B., Anupama, G. V., Nageswara Rao, G. D., & Wani, S. P. (2002). *Socioeconomic characterization and analysis of resource-use patterns is community watersheds in semi-arid India* (Working paper series no. 12). Patancheru 502 324, Andhra Pradesh, India: International Crops Research Institute for the Semi-Arid Tropics, 44 pp.

Singh, P., Alagarswamy, G., Pathak, P., Wani, S. P., Hoogenboom, G., & Virmani, S. M. (1999). Soybean-chickpea rotation on Vertic Inceptisols I. effect of soil depth and landform on light interception, water balance and crop yields. *Field Crops Research, 63*(3), 211–224.

Swindale, L. D. (1982). Distribution and use of arable soils in the semi-arid tropics. In: Managing soil resources., Plenary Session Paper, Trans.12th International Congress of Soil Science, New Delhi, pp. 67–100.

Walker, T. S., Ryan, J. G., Kshirsagar, K. G., & Sarin, R. (1983). *The economics of deep Vertisol technology options: Implication for design testing and evaluation* (Seminar on technology options and economic policy for dryland agriculture). Patancheru: ICRISAT.

Wani, S. P., Sreedevi, T. K., Pathak, P., Singh, P., & Singh, H. P. (2001a). *Integrated watershed management through a consortium approach for sustaining productivity of rainfed areas: Adarsha watershed, Kothapally, India, Andhra: A case study*. Paper presented at Brainstorming workshop on policy and institutional options for sustainable management of watersheds, 1–2 November 2001, ICRISAT, Patancheru 502 324, Andhra Pradesh, India.

Wani, S. P., Pathak, P., Tan, H. M., Ramakrishna, A., Singh, P., & Sreedevi, T. K. (2001b). Integrated watershed management for minimizing land degradation and sustaining productivity in Asia. In Z. Adeel (Ed.), *Integrated land management productivity in Asia* (pp. 207–230). Proceedings of Joint UNU-CAS International Workshop, Beijing, China, 8–13 September 2001.

Wani, S. P., Sreedevi, T. K., Singh, H. P., & Pathak, P. (2002a). *Farmer participatory integrated watershed management model: Adarsha Watershed, Kothapally, India. A success story*. Patancheru: International Crops Research Institute for the Semi-Arid Tropics, 22 pp.

Wani, S. P., Rego, T. J., & Pathak. (eds.). (2002b). Improving management of natural resources for sustainable rainfed agriculture. In *Proceedings of the training workshop on On-farm Participatory Research Methodology, 26–31 July 2001, Khon Kaen, Bangkok, Thailand*. Patancheru: International Crops Research Institute for the Semi-Arid Tropics , 68 pp.

Wani, S. P., Pathak, P., Jangawad, L. S., Eswaran, H., & Singh, P. (2003a). Improved management of Vertisols in the semiarid tropics for increased productivity and soil carbon sequestration. *Soil Use and Management, 19*(3), 217–222. ISSN 1475-2743.

Wani, S. P., Singh, H. P., Sreedevi, T. K., Pathak, P. Rego, T. J., Shiferaw B., & Iyer, S. R. (2003b). *Farmer participatory integrated watershed management: Adarsha Watershed, Kothapally, India*. An innovative and upscalable approach INRM case study 7.

Chapter 3
Climate Change Impacts at Benchmark Watershed

A. V. R. Kesava Rao, S. P. Wani, and K. Srinivas

Abstract Knowledge of climate and weather helps in devising suitable strategies and managing crops to take advantage of the favourable weather conditions and minimizing risks due to adverse weather conditions. The role of climate assumes greater importance in the semi-arid rainfed regions where moisture regime during the cropping season is strongly dependent on the quantum and distribution of rainfall vis-à-vis the soil water holding capacity and water release characteristics. Evidences over the past few decades show that significant changes in climate are taking place all over the world as a result of enhanced human activities through deforestation, emission of various greenhouse gases (GHGs) and indiscriminate use of fossil fuels. Various studies show that climate change in India is real and it is one of the major challenges faced by Indian agriculture. Agroclimatic analyses of the watersheds based on long-term weather data include concepts of rainfall probability, dry and wet spells, water balance, length of growing period (LGP), occurrence of droughts, climate variability and projected climate change. Long-term weather data of Kothapally watershed was obtained from installed automatic weather station and Indian Meteorological Department (IMD) gridded data and analysed for characterizing the agroclimate and assessing the climate variability. Results indicate clear increasing trends in temperature and considerable changes in rainfall. Climate projections also indicate large change in temperature and rainfall at Kothapally in the future. Implementing Integrated Watershed Management Programme in a holistic way can mitigate the adverse effects of climate variability and change and enhance the capacity of small farm holders to manage extremes of drought and floods in a sustainable way.

Keywords Rainfed LGP · Gridded climate data · Projected climate change

A. V. R. Kesava Rao (✉) · K. Srinivas
ICRISAT Development Center, International Crops Research Institute
for the Semi-Arid Tropics (ICRISAT), Hyderabad, Telangana, India
e-mail: k.rao@cgiar.org

S. P. Wani
Former Director, Research Program Asia and ICRISAT Development Centre, International
Crops Research Institute for the Semi-Arid Tropics (ICRISAT), Hyderabad, Telangana, India

© Springer Nature Switzerland AG 2020
S. P. Wani, K. V. Raju (eds.), *Community and Climate Resilience in the Semi-Arid Tropics*, https://doi.org/10.1007/978-3-030-29918-7_3

3.1   Introduction

With the ever-increasing need for food, shelter and energy, maximizing the agricultural production from rainfed areas in a sustainable manner has become the most important aspect of agricultural research. The saying that "farmers learn to live with the limitations of their local climatic conditions through trial and error over generations" is no more wholly true. Past experience provides them with diverse information on rainfall, floods, droughts, etc. Yet for modern agriculture, this is not enough. It is now clear that for deriving the maximum and sustained agricultural yield from watersheds, farmers should have access to proper knowledge of the prevailing agroclimatic conditions.

Weather, the day-to-day state of atmosphere, consists of short-term variation of energy and mass exchanges within the atmosphere and between the earth and the atmosphere. It results from processes attempting to equalize differences in the distribution of net radiant energy from the sun. Acting over an extended period of time, these exchange processes accumulate to become *climate*. In simple terms, climate is the synthesis of weather at a given location over a period of about 30 years. Climate, therefore, refers to the characteristic condition of the atmosphere deduced from repeated observations over a long period. More than a statistical average, climate is an aggregate of environmental conditions involving heat, moisture and air movement. Any study of climate must consider extremes in addition to means, trends, fluctuations, probabilities and their variations in time and space.

The full potential of climate as an agricultural resource has not been used or ever realized. As a result, several crops are grown traditionally in areas without any knowledge of the suitability of climate. Thus, on the one hand, poor yields of crops are obtained, and on the other, much of the production potential of this vast resource is left unutilized. It is impossible to tame the weather on a large scale, or even be in complete harmony with it. However, it is inevitable to make adjustment with the weather or harness the maximum benefit from this resource. In this context, knowledge on agroclimatology of a region is a valuable tool in crop planning.

The importance of climate assumes greater importance in rainfed regions where moisture regime during the cropping season is highly variable and is strongly dependent on the quantum and distribution of rainfall vis-à-vis the soil water holding capacity (WHC) and water release characteristics. Even in irrigated agriculture, where manipulation of moisture regime alone is possible, the thermal and radiation regimes influence the choice of crops, cropping patterns and the optimum dates of sowing for achieving better crop yields. In addition, weather abnormalities like cyclones, floods, droughts, hailstorms, frost, high winds and extreme temperatures (low and high) will lead to natural disasters affecting agricultural productivity. A thorough understanding of the climatic conditions will help in devising suitable management practices for taking advantage of the favourable weather conditions and avoiding or minimizing risks due to adverse weather conditions.

3.2 Kothapally Watershed

Adarsha watershed at Kothapally is located at 17.375 °N latitude, 78.119 °E longitude and about 612 m above mean sea level (msl) in the Shankarpally Mandal, Ranga Reddy District of undivided Andhra Pradesh, India. Kothapally falls under hot moist semi-arid agroecological subregion (AESR) with deep loamy and clayey mixed deep black soils (Vertisols) and associated shallow depth black soils. Available water capacity of these soils is medium to very high (100–300 mm).

3.3 Data and Methods

3.3.1 Automatic Weather Station

In the year 1999, ICRISAT established an automatic weather station (AWS) at Adarsha Watershed, Kothapally. Objectives are to collect reliable weather data for agricultural research and enhance the awareness of weather and climate among the farming community and school children. The AWS was initially installed on farmer's field, and later in 2008 it was shifted to the Zilla Parishad High School for providing better access to the students and effective dissemination of weather information to the farmers. This weather station records rainfall, air temperature, relative humidity, solar radiation, wind speed, wind direction and soil temperatures at 5, 10, 20, 30 and 50 cm depths. The sensors are scanned at 1 min interval and data are permanently stored as hourly averages or totals. In addition, daily summaries are calculated at 0830 in the morning for past 24 h. A GPRS communication system is established at ICRISAT to download the data from AWS on hourly basis. Tabular reports are generated daily and data are graphically printed at weekly/monthly intervals. Rainfall data for the year 1998 was collected from the nearby rain gauge station at Shankarpally; this rain gauge is maintained by the state government. Other weather data for the year 1998 was collected from the nearby weather station maintained by ICRISAT. Thus, a total weather data for 20 years was used for the agroclimatic characterization of the Adarsha watershed, Kothapally.

3.3.2 IMD Gridded Climate Data

A high-resolution daily gridded temperature data set for the Indian region was developed using temperature data of 395 quality controlled stations for the period 1969–2005. A modified version of the Shepard's angular distance weighting algorithm was used for interpolating the station temperature data into 1° latitude × 1° longitude grids (Srivastava et al. 2009). Using cross validation, errors were estimated and found less than 0.5 °C. The data set was also compared with another high-resolution data set and found comparable. The gridded temperature data were updated by IMD for the period 1951–2016; these updated data were procured from the IMD, and daily gridded maximum and minimum temperature data of representative pixel for Kothapally area retrieved. A new daily gridded rainfall data (IMD 4) set at a high spatial resolution ($0.25 \times 0.25°$, latitude $\times$ longitude) covering a long period of 110 years (1901–2010) over the Indian mainland was developed based on inverse distance weighted interpolation (Pai et al. 2014). The gridded data set was developed after making quality control of basic rain gauge stations. The comparison of IMD 4 with other data sets suggested that the climatological and variability features of rainfall over India derived from IMD 4 were comparable with the existing gridded daily rainfall data sets. In addition, the spatial rainfall distribution, like heavy rainfall areas in the orographic regions of the west coast and over northeast, low rainfall in the leeward side of the Western Ghats, etc., was more realistic and better presented in IMD 4 due to its higher spatial resolution and to the higher density of rainfall stations used for its development. The IMD 4 data were updated by IMD to include data up to the year 2016; this updated data were procured from the IMD, and daily gridded rainfall data of representative pixel for Kothapally area for 56 years (1961–2016) was retrieved. These gridded data were used for understanding the temperature and rainfall trends.

Agroclimatic characterization of Kothapally watershed was carried out following well-known and popular methods used earlier (Kesava Rao et al. 2007, 2008). Incomplete gamma and Markov chain methods were used for studying rainfall probabilities. Characterization of a watershed based on average rainfall can yield good results, provided the rainfall distribution is normal. However, in the semi-arid tropics, weekly rainfall totals include a number of zeros. Hence, several researchers suggested the fitting of incomplete gamma distribution to this kind of skewed data (Thom 1958; Krishnan and Kushwaha 1972; Khambete and Biswas 1978; Biswas and Khambete 1979; Biswas and Basarkar 1982). Weekly rainfall that can be expected at different probability levels was computed based on incomplete gamma distribution model. Initial and conditional probabilities of receiving different amounts of rainfall were computed following the method of Virmani et al. (1982).

Modified FAO-Penman-Monteith method (Allen et al. 1998) was used to estimate potential evapotranspiration. Plant extractable water was estimated from soil characteristics. Modified Thornthwaite and Mather method (1955) was used to compute water balances and classification of climate. Length of growing period (LGP) and dry and wet spells during the crop growth period are calculated based on index of moisture adequacy (IMA).

3.3.3　Droughts

According to India Meteorological Department, meteorological drought over an area is defined as a situation when the rainfall received over the area is less than 75% of its long-term average value. It is further classified as "moderate drought" if the rainfall deficit is between 26% and 50% and as "severe drought" when the deficit exceeds 50% of the normal. This criterion was used to classify droughts at Kothapally. For classifying drought, 30-year normal is needed, and hence, in this study gridded rainfall data for 1931–1960 is taken for computing weekly normals for Kothapally. Actual rainfall of each year for the period 1961–2016 was compared with the normal and each year drought if any was classified. Meteorological drought conditions occurring for more than two consecutive weeks certainly lead to agricultural drought and thus occurrence of agricultural droughts with duration of more 2 weeks, 3 weeks and 4 weeks was identified for each year.

3.3.4 Projected Climate Change

To understand the future climatic characters of Kothapally, climate projections as described by the climate projection model CESM1-CAM5 for the year 2030 and for the RCP 8.5 were considered because this is one of the models which best capture the pattern in Indian summer monsoon rainfall over the historical period (Jena et al. 2015).

3.4 Results and Discussion

Analysis of the weather data from the automatic weather station showed that average annual rainfall for Kothapally is about 826 mm and highest one-day rainfall of 346 mm occurred on 24 August 2000. Normal climatic characters of Kothapally are shown in Table 3.1. Normal date of onset of southwest monsoon over Kothapally is around 5 June and the monsoon withdraws by the first week of November.

Table 3.1 Climatic characters of Kothapally

Character	Jan	Feb	Mar	Apr	May	Jun	Jul	Aug	Sep	Oct	Nov	Dec
Rainfall – average (mm)	4	6	20	28	44	112	150	206	156	88	10	2
Rainfall – std. dev. (mm)	7	16	35	30	54	59	83	104	84	76	12	3
Rainfall – CV (%)	218	252	172	108	125	54	56	51	54	86	125	243
Rainy days	0	0	2	2	3	7	10	11	9	5	1	0
Max T (°C)	29.4	31.9	35.2	37.8	38.7	34.0	30.8	29.5	30.0	30.7	29.7	29.0
Min T (°C)	13.5	15.9	19.0	22.7	24.7	23.4	22.3	21.7	21.3	19.2	15.2	12.1
RH – morning (%)	85	76	66	62	60	80	86	90	92	90	89	87
RH – afternoon (%)	29	25	21	21	24	43	56	62	60	47	38	30
Wind speed (km/h)	3.9	4.5	5.2	5.6	6.8	7.8	7.9	6.9	5.7	4.3	3.9	3.6
Solar radiation (MJ /sq. m)	16.0	18.3	19.9	21.2	21.6	17.2	14.9	14.6	16.0	16.5	15.9	15.6

About 624 mm of rainfall is received in the southwest monsoon season which is about 75% of the annual rainfall. Monthly rainfall distribution at Kothapally indicates that August is the rainiest month of the year with an average rainfall of about 206 mm followed by September with about 156 mm of rainfall. December–February is winter period, which is dry having a low rainfall of about 12 mm and with high coefficient of variation of rainfall. During the 4 months, June–September (southwest monsoon period), rainfall variability is less but still above 50%. Rainy day is defined as a day which receives above 2.4 mm, and it is seen that there are only 50 rainy days at Kothapally, of which 30 rainy days are present during July to September, indicating the prime rainfed crop-growing period.

Climatic water balance was worked out based on weekly rainfall and potential evapotranspiration (PET) and shown in Fig. 3.1 for the rainfed crop-growing period. Rainfall expected at 50 and 75% probability levels is also shown. Above 30 mm rainfall can be expected at 50% probability in just 4 weeks during the period. At 75% probability (in 3 out of 4 years), weekly rainfall is never above 20 mm, and after middle of October, drastic reduction in rainfall is observed.

Annual rainfall is about 826 mm, while the annual PET is about 1655 mm indicating that atmospheric and crop water requirements are very high compared to rainfall; only for about 14 weeks rainfall exceeds PET starting from the second half of July. Average annual water surplus is about 155 mm and average annual water deficit is about 987 mm. Second half of August in general receives high rainfall, and this is the time for water harvesting for use in the end period of *kharif* (rainy season) crops and for *rabi* (post-rainy season) crops.

At Kothapally, the initial probabilities of receiving a rainfall of more than 30 mm per week are generally moderate throughout the period during July second week to last week of October, indicating the possibility of little moisture stress during the

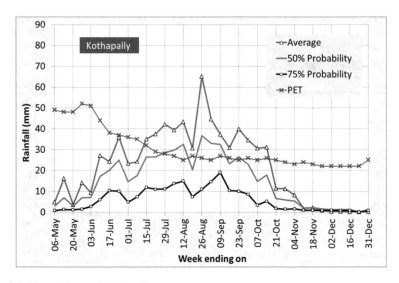

Fig. 3.1 Water balance of Kothapally

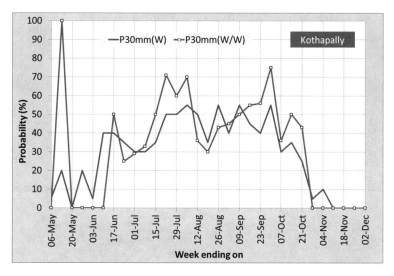

Fig. 3.2 Initial and conditional probability rainfall at Kothapally

crop-growing period (Fig. 3.2). The conditional probability line indicates that there is high probability of more than 60% to receive 30 mm of rainfall during third week of July to first week of August, provided the respective previous weeks are wet with a rainfall of above 30 mm. During the crop-growing period, last week of September has above 70% probability of receiving at least 30 mm of rainfall provided the third week of September received above 30 mm of rainfall. This kind of information will be of great practical use in planning water harvesting and storing in Kothapally watershed.

Knowledge on the amount of rainfall and rainfall intensity helps in understanding the water harvesting potential at a watershed. Daily rainfall data of Kothapally for the 20 years were classified into four categories having days receiving rainfall of more than 2.5, 10, 25 and 50 mm in each month and are presented in Fig. 3.3.

It is observed that on an average, there are about 50 days in a year receiving a rainfall above 2.5 mm per day. There is large variation across years; about 74 days were there in the year 2010 and only 40 days were observed in the year 2012. In about 7 days, rainfall in a day could be above 25 mm indicating water harvesting potential. Rainfall between 2.5 and 10 mm per day is ideal for meeting the crop water requirements and enhancing the soil moisture storage; such days are about 25 in a year at Kothapally.

3.4.1 Length of Rainfed Crop-Growing Period (LGP)

Knowledge on the date of onset of rains will help to plan the agricultural operations better, particularly, land preparation and sowing. Length of rainfed crop-growing period is the period of the year in which crops could be grown successfully as both

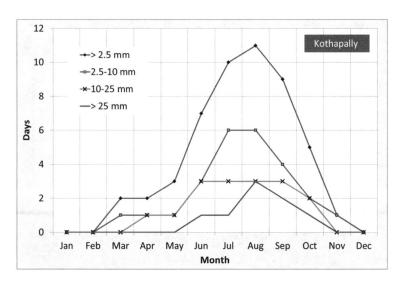

Fig. 3.3 Monthly rainfall intensities at Kothapally

rainfall and moisture stored in the soil will meet the moisture demands of crops. Therefore, the LGP depends not only on the rainfall distribution but also on the type of soil, soil depth, water retention and release characteristics of the soil. This assumes greater importance from a watershed perspective where soil depth in a topo-sequence can also alter the LGP across the watershed being highest in the low-lying regions and lowest in the upper reaches of the watersheds. Several methods were used for estimating the LGP using rainfall (Ashok Raj 1979, IMD 1991; Sivakumar et al. 1993). The National Bureau of Soil Survey and Land Use Planning (Velayutham et al. 1999) estimated LGP using the PET and rainfall. Using the water balance, week-wise index of moisture adequacy (IMA) was computed, which is defined as the ratio of the actual evapotranspiration to the potential evapotranspiration expressed as a percentage. The beginning and end of the growing season were identified based on the IMA. The growing season begins when the IMA is above 50% consecutively for at least 2 weeks, starting from the middle of May. The end of the season was identified when the IMA falls below 25% for two consecutive weeks, when seen backwards starting from the end of January. LGP was computed for the 20 years and the variability is shown in Fig. 3.4.

Crop-growing period at Kothapally can begin as early as last week of May but could be delayed to as late as third week of August, a difference of about 80 days. Crop-growing period can end as early as end of November, but could be extended up to middle of January, a difference of about 40 days. Thus it is seen that more variability exists in the beginning of the crop-growing season compared to the end of season. On an average the growing period begins by the first week of July and ends by first week of December, making the length of the growing period as about

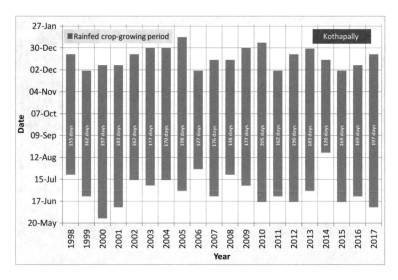

Fig. 3.4 Variability in rainfed crop-growing period at Kothapally

140–160 days. Thus, there is more risk in the beginning of the season and highlights the need for weather-based sowing advisories for the farmers in the region based on seasonal climate and medium-range weather forecasts.

3.4.2 Temperature and Rainfall Trends

Global warming due to increase in the carbon dioxide in the atmosphere is witnessed in several parts of India including Kothapally. Sixty-six years (1951–2016) of IMD gridded data were used to assess the changes in temperature at Kothapally (Fig. 3.5).

Analysis of temporal variations of temperature during *kharif* (Jun–Oct) and *rabi* (Nov–Feb) seasons shows that there appears to be a clear increase in maximum temperature of about 1 °C in *kharif*, while the increase in *rabi* is higher at about 1.4 °C. This indicates that in future, if these trends continue, Kothapally area is likely to have higher temperatures during both *kharif* and *rabi* seasons, impacting the crop duration and productivity.

To understand the rainfall changes, 56 years (1961–2016) of monthly rainfall data of Kothapally was used. The 30-year period from 1961 to 1990 was considered as normal and rainfall for this period was compared with the average rainfall of 26 years (1991–2016) and results are shown in Fig. 3.6. It is seen that though the net change is positive at about 26 mm annually, three important months of the crop-growing period, i.e., June, July and September, have shown reduced rainfall of about 10, 15 and 20 mm, respectively.

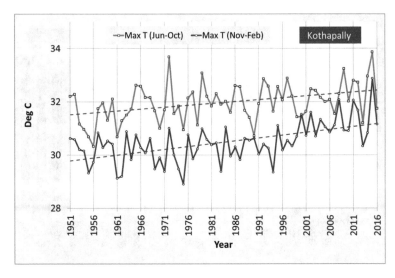

Fig. 3.5 Temperature trends at Kothapally

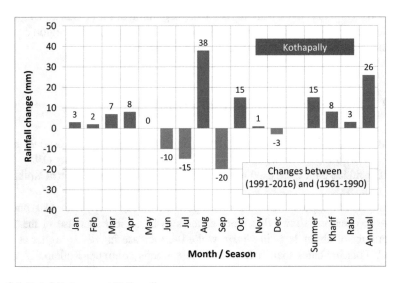

Fig. 3.6 Rainfall changes at Kothapally

Frequency of droughts with 2, 3 and 4 weeks duration was worked out based on the criterion of India Meteorological Department (Fig. 3.7).

It is observed that occurrence of drought during the crop-growing period is common, and in general, almost 30% of the time drought of 2-week duration occurred

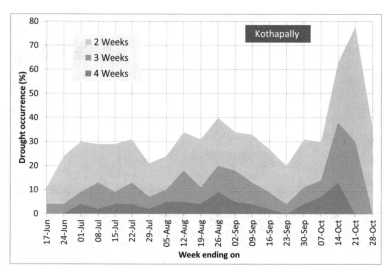

Fig. 3.7 Drought occurrence at Kothapally

in the past. Drought of 4-week duration is disastrous and there is only 10% chance for this kind of drought to occur. Middle of October has higher tendency for this kind of drought and this period coincides with crop maturity stage highlighting the need for proper water management at watershed level during this period.

3.4.3 Climate Change at Kothapally

Geographical boundaries of climatic zones vary with time when changes in temperature and rainfall become considerable. Climate change is a major issue for sustainable agriculture, and changes in temperature and rainfall will ultimately change the type of climate a region enjoys. A study carried out by ICRISAT under the National Initiative on Climate Resilient Agriculture (NICRA) project based on the gridded rainfall and temperature data of India Meteorological Department quantified the changes in areas under different climates in India. The study indicated a net reduction in the dry sub-humid area (10.7 m ha) in the country, of which about 5.1 million ha (47%) shifted towards the drier side and about 5.6 million ha (53%) became wetter, comparing the periods 1971–1990 and 1991–2004 (Kesava Rao et al. 2013). Using the climatic water balance approach and classification of climates (Thornthwaite and Mather 1955), climate of Kothapally was classified for 20 years (1998–2017) and is shown in Fig. 3.8.

Normal climate of Kothapally is semi-arid and it is seen that the climate was mostly under this type; once the climate shifted to dry subhumid (year 1988) and once to arid (year 2011) type of climate. Kothapally showed a slight tendency towards drier climate till 2014.

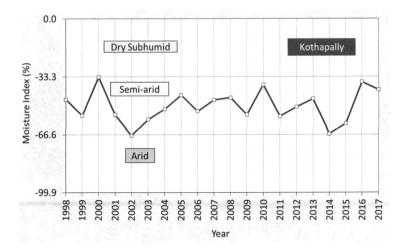

Fig. 3.8 Changes in climate type at Kothapally

Table 3.2 Temperature and rainfall projections for Kothapally

Element	Jan	Feb	Mar	Apr	May	Jun	Jul	Aug	Sep	Oct	Nov	Dec	Year
Max T °C	0.1	0.4	0.7	1.2	0.8	1.2	1.1	1.0	0.7	1.0	1.5	0.8	0.9
Min T °C	1.4	0.8	0.9	0.7	0.6	0.9	1.0	0.9	0.9	1.3	2.1	2.2	1.1
Rainfall (mm)	4	4	4	0	16	4	−13	−13	5	27	6	21	65

3.4.4 Projected Changes in Temperature and Rainfall at Kothapally

Climate change is one of the major challenges faced by rainfed agriculture in the semi-arid tropics (SAT). Global atmospheric concentration of CO_2 has increased from pre-industrial level of 280 parts per million (ppm) to 400 ppm in 2014 and reached to about 407.75 ppm by April 2017. With global climate change, rainfall variability is likely to further increase. When decrease in rainfall is coupled with higher water requirements due to elevated temperatures, rainfed crop-growing period is likely to shorten.

The earth system model CESM1 (version 1.2.1) is a fully coupled global climate model. Many physics-based models serve for the different earth system components. The atmosphere component, CAM5, provides a set of physics parametrizations, and several dynamical cores, which also include advection (Baumgaertner et al. 2016).

In the present study, the CESM1.1_CAM5 model for the Coupled Model Intercomparison Project phase 5 (CMIP5) is considered, and rainfall and temperature projections for the year 2030 for the RCP 8.5 (Representative Concentration Pathways) based on this model were used to understand the future climate conditions likely at Kothapally region (Table 3.2).

Annual maximum temperature is expected to be higher by about 0.9 °C, and annual minimum temperature is likely to be higher by about 1.1 °C. Maximum temperatures during April, June and November could be higher by about 1.2–1.5 °C, while minimum temperatures during November to December are likely to increase by 2.1–2.2 °C. Annual rainfall is likely to increase by about 65 mm. Rainfall during July and August is likely to be lower by about 13 mm. July and August are crop grand growing months, and any reduction in rainfall in these months is likely to have adverse effect on crop growth, phenology and productivity. October is likely to receive more rainfall of about 27, which is likely to offset the low rainfall situation in the middle of the crop-growing period and help water harvesting and storage for use in the *rabi* season. However, higher rainfall leads to flooding, inundation, runoff and soil loss, which needs to be addressed.

3.5 Conclusions

Agroclimatic analysis at watershed level coupled with crop-simulation models, and better seasonal and medium duration weather forecasts, helps build resilience to climate variability/change. Farmers having access to climate and weather information are more likely to take better crop management actions. The automatic weather station installed at the Kothapally watershed has created awareness among the community. Support was given to local farmers for planning pesticide sprays and scheduling irrigations. Climate awareness programmes were conducted for the school students. Training was provided to four girls and two boys (from VIII and IX standard classes) of the school to operate the data logger and to display daily weather data in the school and outside the school campus for farmers use. Historical weather data analysis of Kothapally indicated clear trends in temperature and considerable changes in rainfall. Climate projections also indicate large change in temperature and rainfall at Kothapally in the future.

Resilience to climate change requires identifying climate-resilient crops adaptation practices and enhancing degree of awareness of community. ICRISAT's research findings showed that integrated genetic and natural resources management (IGNRM) through participatory watershed management is the key for improving rural livelihoods in the SAT (Wani et al. 2002, 2003; Wani and Rockström 2011). Even under a climate change regime, crop yield gaps can still be significantly narrowed down with improved management practices and using germplasm adapted for warmer temperatures (Wani et al. 2003, 2009; Cooper et al. 2009). Integrated watershed management (IWM) comprises improvement of land and water management, integrated nutrient management including application of micronutrients, improved varieties and integrated pest and disease management for substantial productivity gains and economic returns by farmers. The goal of watershed management is to improve livelihood security by mitigating the negative effects of climatic variability while protecting or enhancing the sustainability of the environment and the agricultural resource base. Scaling up of issue of weather-based agroadvisories for better crop management using new ICT tools to reach the farming community will enhance resilience to climate variability and change.

References

Allen, R. G., Pereria, L. S., Raes, D., & Smith, M. (1998). *Crop evapotranspiration – Guidelines for computing crop water requirements* (FAO Irrigation and Drainage Paper 56). Rome: FAO.

Ashok Raj, P. C. (1979). *Onset of effective monsoon and critical dry spells* (IARI research bulletin no. 11). New Delhi: Water Technology Centre, IARI.

Baumgaertner, A. J. G., Jöckel, P., Kerkweg, A., Sander, R., & Tost, H. (2016). Implementation of the Community Earth System Model (CESM) version 1.2.1 as a new base model into version 2.50 of the MESSy framework. *Geoscientific Model Development, 9*, 125–135.

Biswas, B. C and Basarkar, S. S. (1982). Weekly rainfall probability over dry farming tract of Gujarat. Annals of Arid Zone Vol. 21 (3).

Biswas, B. C., & Khambete, N. N. (1979). Distribution of short period rainfall over dry farming tract of Maharashtra. *Journal of Maharashtra Agricultural Universities, 4*(2), 187.

Cooper, P., Rao, K. P. C., Singh, P., Dimes, J., Traore, P. S., Rao, A. V. R. K., Dixit, P., & Twomlow, S. J. (2009). *Farming with current and future climate risk: Advancing a "hypothesis of hope" for rainfed agriculture in the semi-arid tropics* (Journal of SAT Agricultural Research 7). Patancheru: An Open Access Journal Published by ICRISAT.

India Meteorological Department (IMD). (1991). *Sowing rains over Madhya Pradesh, climatological characteristics and agricultural importance* (Meteorological Monograph Agrimet/ No.12/1990) (33 pp). Pune: Investigation and Development Unit, Office of the Additional Director General of Meteorology (Research).

Jena, P., Azad, S., & Rajeevan, M. N. (2015). Statistical selection of the optimum models in the CMIP5 dataset for climate change projections of Indian monsoon rainfall. *Climate, 2015*(3), 858–875. https://doi.org/10.3390/cli3040858.

Kesava Rao, A. V. R, Wani, S. P., Piara, S., Irshad Ahmed, M., & Srinivas, K. (2007). Agroclimatic characterization of APRLP-ICRISAT nucleus watersheds in Nalgonda, Mahabubnagar and Kurnool districts. *Journal of SAT Agricultural Research..* Patancheru 502 324, Andhra Pradesh, India: International Crops Research Institute for the Semi-Arid Tropics (ICRISAT). *3*(1), 55 pp.

Kesava Rao, A. V. R., Wani, S. P., Singh, P., Rao, G. G. S. N., Rathore, L. S., & Sreedevi, T. K. (2008). Agroclimatic assessment of watersheds for crop planning and water harvesting. *Journal of Agrometeorology, 10*(1, June), 1–8.

Kesava Rao, A. V. R., Suhas Wani, P., Singh, K. K., Irshad Ahmed, M., Srinivas, K., Snehal Bairagi, D., & Ramadevi, O. (2013). Increased arid and semi-arid areas in India with associated shifts during 1971-2004. *Journal of Agrometeorology, 15*(1, June), 11–18.

Khambete, N. N., & Biswas, B. C. (1978). Characteristics of short period rainfall in Gujarat. *Indian Journal of Meteorology, Hydrology & Geophysics, 29*(3), 521.

Krishnan, A., & Kushwaha, R. S. (1972). Mathematical distribution of rainfall in arid and semi-arid zones of Rajasthan. *Indian Journal of Meteorology & Geophysics, 23*(2), 153–160.

Pai, D. S., Sridhar, L., Rajeevan, M., Sreejith, O. P., Satbhai, N. S., & Mukhopadyay, B. (2014). Development of a new high spatial resolution (0.25° × 0.25°) long period (1901–2010) daily gridded rainfall data set over India and its comparison with existing data sets over the region. *Mausam, 65*(1), 1–18.

Sivakumar, M. V. K., Maidoukia, A., & Stern, R. D. (1993). *Agroclimatology of West Africa: Niger* (2nd ed.). ((In En. Fr. Summaries in En. Fr.) Information bulletin no. 5. Patancheru: ICRISAT, 108 pp.

Srivastava, A. K., Rajeevan, M., & Kshirsagar, S. R. (2009). Development of a high resolution daily gridded temperature data set (1969–2005) for the Indian region. *Atmospheric Science Letters.* 2009, https://doi.org/10.1002/a.s.l.232.

Thom, H. C. S. (1958). A note of Gamma distribution. *Monthly Weather Review, 86*(4), 117.

Thornthwaite, C. W., & Mather, J. R. (1955). *The water balance* (Publications in climatology, Vol. VIII. no. 1). Centerton: Drexel Institute of Technology, Laboratory of Climatology.

Velayutham, M., Mandal, D. K., Mandal, C., & Sehgal, J. (1999). *Agro-ecological sub regions of India for development and planning* (NBSS Publication 35) (452 pp). Nagpur: National Bureau of Soil Survey and Land Use Planning.

Virmani, S. M., Siva Kumar, M. V. K., & Reddy, S. J. (1982). *Rainfall probability estimates for selected locations of semi-arid India* (Research bulletin no. 1) (170 pp). Andhra Pradesh: International Crops Research Institute for the Semi-Arid Tropics.

Wani, S. P., & Rockström, J. (2011). Watershed development as a growth engine for sustainable development of dryland areas. In P. Wani Suhas, J. Rockström, & K. L. Sahrawat (Eds.), *Integrated watershed Management in Rainfed Agriculture* (pp. 35–52). Boca Raton: CRC Press.

Wani, S. P., Rego, T. J, Pathak, P., & Piara, S. (2002). Integrated watershed management for sustaining natural resources in the SAT. In *Proceedings of International Conference on Hydrology and Watershed Management, 18–20 December 2002, Hyderabad, India* (pp. 227–236). Hyderabad: Jawaharlal Nehru Technological University (JNTU).

Wani, S. P., Singh, H. P., Sreedevi, T. K., Pathak, P., Rego, T. J., Shiferaw, B., & Rama, I. S. (2003). Farmer-participatory integrated watershed management: Adarsha watershed, Kothapally India, an innovative and upscalable approach. A case study. In R. R. Harwood & A. H. Kassam (Eds.), *Research towards integrated natural resources management: Examples of research problems, approaches and partnerships in action in the CGIAR* (pp. 123–147). Washington, DC: Interim Science Council, Consultative Group on International Agricultural Research.

Wani, S. P., Sreedevi, T. K., Rockström, J., & Ramakrishna, Y. S. (2009). Rainfed agriculture – Past trend and future prospects. In S. P. Wani, J. Rockström, & T. Oweis (Eds.), *Rainfed agriculture: Unlocking the potential* (Comprehensive assessment of water management in agriculture series) (pp. 1–35). Wallingford: CAB International.

Chapter 4
Soil management for Sustained and Higher Productivity in the Adarsha Watershed

Girish Chander, S. P. Wani, Raghavendra Sudi, G. Pardhasaradhi, and P. Pathak

Abstract Kothapally watershed is a typical representative of rain-fed (800 mm rainfall) semi-arid tropics (SAT) with varying soil depth in the watershed and widespread soil degradation as the major challenge coupled with low crop yields and family incomes. Before the onset of initiative during 1999, soil health mapping and baseline surveys showed varying soil depth in fields at different topo-sequence, macro-/micronutrient deficiencies along with low soil carbon (C) levels and heavy soil loss through erosion that compromised with crop production in the watershed. Inappropriate fertilizer management decisions leading to negative budget for primary nutrients in major crops/cropping systems highlighted suboptimal fertilizer use. Unawareness about micro-/secondary nutrient deficiencies like sulphur (S), boron (B) and zinc (Zn) and lack of addition of such fertilizers contributed to low crop yields and declining fertilizer and water use efficiency. Farmers participatory trials highlighted yield loss of 13–39% in crops like sorghum and maize in the absence of deficient micro-/secondary nutrient fertilizers. Recycling of on-farm wastes through vermicomposting and biomass generation using N-rich *Gliricidia* on farm boundaries were promoted for fertilizer savings and crop yield benefit alongside soil carbon building for developing resilience. The impact of integrated soil health management practices cumulatively observed over 13 years was demonstrated during 2012 soil health mapping that showed improved mean level of soil organic C; available nutrients, viz. phosphorus (P), B, Zn and S; and significantly reduced number of fields with low nutrient/C levels. Along with yield advantage, soil loss was significantly reduced from 3.48 t ha^{-1} in untreated area to 1.62 t ha^{-1} in treated watershed area.

G. Chander (✉) · R. Sudi · G. Pardhasaradhi · P. Pathak
ICRISAT Development Centre, Research Program Asia, International Crops Research Institute for the Semi-Arid Tropics (ICRISAT), Hyderabad, Telangana, India
e-mail: g.chander@cgiar.org

S. P. Wani
Former Director, Research Program Asia and ICRISAT Development Centre, International Crops Research Institute for the Semi-Arid Tropics (ICRISAT), Hyderabad, Telangana, India

© Springer Nature Switzerland AG 2020
S. P. Wani, K. V. Raju (eds.), *Community and Climate Resilience in the Semi-Arid Tropics*, https://doi.org/10.1007/978-3-030-29918-7_4

Keywords Land degradation · Production system · Soil conservation · Soil
organic carbon · Water quality

4.1 Characterization of Soil Health and Issues in Kothapally Watershed

Kothapally watershed represents dryland areas which are categorized by varying
soil depth, prone to severe land degradation, erratic rainfall, high soil erosion, inherently less fertile soils and low rainwater use efficiency. Farmers in dryland areas, in
general, are poor, and their ability to take risk and invest in necessary inputs for
optimizing production is low (Joshi et al. 1996). Watershed programmes in India
therefore are instrumental and silently revolutionalizing the rain-fed agriculture for
improving the productivity in dryland areas with major focus on natural resource
conservation (Wani and Garg 2015; Wani and Patil 2018). Maintaining proper soil
health is one of the essential elements of sustainable agriculture and safeguarding
ecosystem services (Wani et al. 2018). The depletion of soil nutrients often leads to
land degradation and low fertility levels that limit production and reduce water productivity. The impact of land degradation is especially severe on livelihoods of the
poor who heavily depend on natural resources. In case of Kothapally watershed,
with a population of around 1500, around 270 families depend on cultivation and 4
are non-cultivators. The average landholding per household was 1.4 ha (Shiferaw
et al. 2002). Soil health management is not only a prerequisite to strengthen agri-
based enterprises but also a very effective entry point intervention to quickly harness the productivity benefits while bringing on board the majority farmers because
of common interest and benefits to all (Wani et al. 2002, 2009; Dixit et al. 2007;
Chander et al. 2016).

Initial baseline surveys in Kothapally watershed pointed out to poor fertilizer
management practices and declining fertilizer use efficiency. Therefore, to undertake precise diagnosis, representative soil samples were collected from the watershed following the stratified soil sampling method (Sahrawat et al. 2008). Detailed
soil characterization showed low levels of nitrogen (N) (11 mg kg^{-1} soil), phosphorus (P) (1.4–2.2 mg kg^{-1} soil) and micro-/secondary nutrients like sulphur (S), boron
(B) and zinc (Zn) along with low soil organic carbon. Soils are predominantly
Vertisols and associated soils (90%) with dominance of clay (42%, range of
5.16–65.61% across fields). Average composition of other mechanical separates
was 18% (10.21–29.75% range) silt, 24% (8.33–45.71 range) fine sand and 16%
(3.22–43.14 range) coarse sand. The soil depth ranges from 30 to 90 cm and watershed is characterized by an undulating topography with an average slope of about
2.5% (Wani et al. 2003a, b).

These assessments clearly highlighted to focus on promoting need-based sustainable nutrient management including that of micro-/secondary nutrients and soil
C building measures through effectively using on-farm biomass. The soil type and

texture observed also needed broad-bed and furrow (BBF) or conservation furrow (CF) landform systems for addressing the barriers of conveniently taking two crops in a year and storing more soil moisture while reducing runoff. Actually, these Vertisols have poor hydraulic conductivity and consequently are frequently poorly drained. The land management practices like CF at 3–4 m interval or BBF landform system comprising of 1.05 m width raised bed with 0.45 m furrow can effectively address the existing barriers to effectively draining excess water via furrows, enabling land preparation by providing compact furrows to move on while keeping intact the surface bed soil and infiltrating and storing more soil water through intact surface bed soil.

4.2 Nutrient Budgeting of Production Systems

Nutrient budgeting is an important tool in addition to soil health mapping for insight into the balance between inputs and outputs during the crop-growing period. It helps evaluate nutrient management scenarios and identify any production or environmental issues arising out of nutrient excesses or deficits. This technique was adopted in Kothapally watershed for understanding major nutrient-related issues. For this, the watershed was divided into three topo-sequences and nutrient budgets were done using stratified random sampling proportionately for major crops/cropping-system across topo-sequences in both the landforms of flat cultivation (normal practice) as well as broad bed and furrow (BBF, improved practice). The balances showed that all the systems were depleting nitrogen (N) and potassium (K) from soils and that P is applied almost equal to the requirement or more than what is removed by crops (Table 4.1). N, phosphorus (P) and K nutrient uptake was in general greater in the improved BBF system compared to that on the flat landform, apparently because of more crop yield on the BBF landform (Fig. 4.1).

4.3 Scaling Up Balanced Nutrients and Building Soil Carbon

Based on soil analysis results, fertilizer recommendations were discussed with the farmers and promoted for all major crops to take care of incurring soil nutrient deficits. Alongside suboptimal fertilizer use of primary nutrients, the observed deficiencies of micro- and secondary nutrients like B, Zn and S were major constraints for productivity improvement and sustainability. To introduce new practice is always a challenge and collective participatory research and learning is the best way to bring in desired change in current practice. Therefore, farmer participatory research trials were conducted to evaluate micro-/secondary nutrients in crops like sorghum and maize. Amendments with B, alone and in combination, resulted in 13–39% increase in sorghum and maize grain yield (Table 4.2). This tangible benefit was a good trig-

Table 4.1 Soil health status of farmers' fields (263 field samples) in Kothapally watershed after the watershed works, May 2012

	pH	EC (dS m⁻¹) dS/m	Available nutrients (mg kg⁻¹)								Total N	Carbon (%)		
			P	K	S	Zn	B	Fe	Cu	Mn		Organic	Inorganic	Total
% deficient farmers			31	0	37	68	52	0	0	0				
Mean	8.0	0.27	12.1	242	14.1	0.73	0.63	15.15	4.57	12.06	868	1.53	0.41	1.94
Range	6.8–8.6	0.01–0.94	0.3–83.9	55–615	3.1–62.0	0.14–4.69	0.22–1.83	5.6–44.9	1.4–18.4	4.4–33.0	329–1821	0.13–3.11	0.00–1.84	0.54–4.37

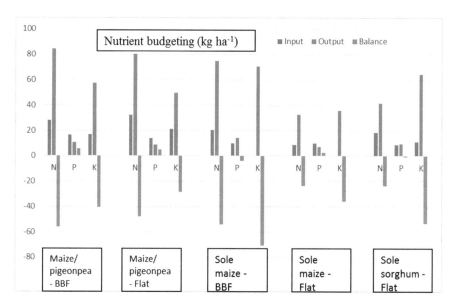

Fig. 4.1 Nutrient budgeting studies in farmers' fields, Adarsha watershed, Kothapally, 1999–2000. (Derived from Wani et al. 2006)

Table 4.2 On-farm (medium soil depth) evaluation of landform management in Adarsha watershed, Kothapally, during 2001

System	Soil type	Landform	Yield (kg ha^{-1})		System productivity (1 + 2)
			Main crop (maize/ sorghum) – 1	Component crop (PP) – 2	(kg ha^{-1})
Maize/PP	Shallow	BBF	1750	380	2130
Maize/PP	Shallow	Flat	1680	290	1970
Maize/PP	Medium	BBF	2830	1070	3900
Maize/PP	Medium	Flat	2780	820	3600

BBF broad bed and furrow, *pp* pigeon pea

ger to adopt use of micro- and secondary nutrient fertilizers by majority farmers in the watershed (Sreedevi et al. 2004; Wani et al. 2006; Dixit et al. 2007) (Fig. 4.2).

In post-green revolution era, fallout of the fertilizer subsidy is that chemical fertilizers are cheaper than organic fertilizers and, so, farmers are tempted to move away from using organic manure for rain-fed agriculture, which is very critical for preserving good soil health (Wani et al. 2016, 2018). Little or no addition of organics coupled with imbalanced use of mineral fertilizers has led to depletion of soil

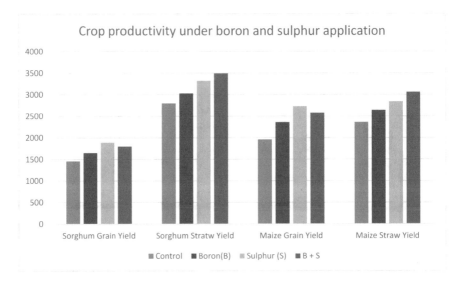

Fig. 4.2 Total productivity of sorghum and maize with boron and sulphur amendments at Adarsha watershed, Kothapally, 2001. (Derived from: Sreedevi et al. 2004; Wani et al. 2006)

organic carbon (C) resulting into its low levels which is one of the major factors for declining soil productivity. Soil organic matter has long been suggested as the single most important indicator of soil productivity (Wani et al. 2003a, 2018). Even small changes in total C content can have large impact on soil biological and physical properties and crop yields.

Recycling large quantities of carbon (C) and nutrients contained in agricultural and domestic wastes (~700 million t organic wastes are generated annually in India) are needed to rejuvenate soil health for enhancing productivity (Nagavallemma et al. 2006; Chander et al. 2013; Wani et al. 2014). Vermicomposting is a simple process of composting with the help of earthworms to produce a better enriched end product. It is one of the easiest methods to recycle organic wastes to produce quality compost for farm requirements (Wani et al. 2014). Vermicompost is, in general, rich in nutrients than other compost due to passage of material through the guts of the worms and gets enriched with nutrients and hormones. Earthworms consume various organic wastes and reduce the volume by 40–60% (Nagavallemma et al. 2006) in 8 weeks after releasing the worms. Vermicompost prepared through decomposing sorghum straw and dung biomass (80:20 ratio, primed with 0.5% urea and 4% rock phosphate) has recorded reasonably high concentration of various nutrients, like 11,100 mg kg^{-1} N, 4300 mg kg $^{-1}$ P, 9600 mg kg $^{-1}$ K, 31500 mg kg^{-1} Ca, 6000 mg kg^{-1} Mg, 17 mg kg^{-1} S, 88 mg kg^{-1} Zn, 17.9 mg kg^{-1} Cu, 7525 mg kg^{-1} Fe, 395 mg kg^{-1} Mn, 91 mg kg^{-1} B and C: N ratio of 11.7 (Chander et al. 2018).

In the background of poor soil health and availability of on-farm biomass, vermicomposting was promoted in the watershed both for field use as well as a microenterprise to generate income through sale (Fig. 4.3). Training was imparted to farmers

Fig. 4.3 Vermicomposting by women self-help groups in Adarsha watershed, Kothapally, in Telangana state (erstwhile undivided Andhra Pradesh), India

and women SHG groups. The raw material used was *Parthenium* (locally known as congress weed) which is an obnoxious invasive weed in the country. The *Parthenium* growing in the village was uprooted by the community through voluntary work for a day in a year and made available to women SHGs for composting. The women group collectively undertook the vermicomposting using the earthworms and enriched with rock phosphate (Nagavallema et al. 2006). Participatory evaluation trials by the farmers showed application of 3 and 5 t ha^{-1} of vermicompost increased tomato yield to 4.8–5.8 t ha^{-1} as compared to the plots (3.5 t ha^{-1}), which received conventional compost. In onion, the application of 2.5 t ha^{-1} of vermicompost + chemical fertilizers gave additional yield of 3.75 t ha^{-1} when compared to fields which received only chemical fertilizers. Similarly, response of vermicompost was recorded for turmeric. It was also observed that the effect of vermicompost was seen even in the next year crops. In addition, for biomass generation, *Gliricidia* plantations were promoted on farm boundaries and N-rich leaves are used in making vermicompost or incorporating in field. Farmers have planted about 50,000 *Gliricidia* saplings on bunds for generating N-rich organic matter in the watershed. On-station watershed studies at ICRISAT have shown that *Gliricidia* loppings provide around 30 kg N ha^{-1} year^{-1} without adversely affecting crop yield (ICRISAT 2002; Wani et al. 2003b).

The impact of integrated soil health management practices continued for 13 years was evident in Kothapally watershed during 2012 soil health mapping. Soil health mapping during 2012 showed improved mean level of soil organic C; available nutrients, viz. phosphorus, B, Zn and S; and significantly reduced number of fields with low nutrient/C levels.

4.4   Conservation of Soil Resources

According to NBSS&LUP, around 146.8 M ha is degraded land in India and water erosion is the major factor (Bhattacharyya et al. 2015). There is annual total soil loss of 5.3 billion tons in India at ~16.4 t ha^{-1} year^{-1} and direct estimated cost of land degradation is around Rs 450 billion equivalent to \$6.4 billion (crop productivity, high-input use, lost nutrients, land use intensity, changing cropping pattern) annually. Watershed management is one of the most trusted and eco-friendly approaches to managing soil loss. In this context, the salient impacts that resulted due to the implementation of this watershed were substantial reductions in runoff and soil loss. Soil and water conservation measures implemented by farmers in individual fields were broad-bed and furrow (BBF) landform, land smoothening, field drains, flat cultivation with conservation furrows and contour planting to conserve in situ soil and water and planting *Gliricidia* on field bunds to strengthen bunds and supply nitrogen (N)-rich organic matter for in situ application to crops. Common wasteland treatment was done by planting *Pongamia pinnata, Jatropha curcas*, custard apple saplings, G*liricidia* saplings and avenue plantation as part of village afforestation programme. The direct benefits of BBF landform were observed over traditional flat landform treatments (Table 4.2). Farmers obtained 250 kg more pigeon pea and 50 kg more maize per hectare using BBF on medium-depth soils than from the flat landform treatment. The farmers with shallow soils reported similar benefits from BBF landform and improved management options for other cropping systems. The BBF system increased the yield of cotton by 32%, pigeon pea by 17%, maize by 25% and sorghum by 21% compared to traditional flat practice. The flat cultivation along with conservation furrow system has increased the yield of cotton by 22%, pigeon pea by 16%, maize by 20% and sorghum by 15% compared to traditional flat practice. The benefits of these in situ soil and water management practices were found better during low- and high-rainfall years. These practices were also found effective in improving soil moisture and controlling runoff, peak runoff rate and soil loss.

Community-based interventions were implemented on common resources like 14 water storage structures (one earthen and 13 masonry) with a capacity of 300–2000 m^3, 97 gully control structures, 60 mini percolation pits, 1 gabion structure (Fig. 4.4) for increasing groundwater recharge, a 500 m long diversion bund and field bunding on 38 ha that were completed. Due to these watershed interventions, the groundwater recharge and its availability increased substantially. Despite of several fold increase in the numbers of borewells, the groundwater levels in the watershed were maintained. Even during the post-rainy and summer seasons, the performance of open wells improved substantially. For example, during the post-rainy season, the average area irrigated by each open well increased from 0.6 to 1.1 ha. The data from 2000 to 2014 clearly show that the watershed interventions resulted big increase in groundwater availability. Increase groundwater availability had led to increased investments as well as better adoption of improved agricultural technology (improved crop varieties, chemical fertilizers, drip irrigation, cultivation

Fig. 4.4 Water storage structure in Adarsha Watershed, Kothapally, in Telangana state, India

of high-value crops and others) by watershed farmers. It has contributed in increasing agricultural productivity and income as well as in increasing cropping intensity and crop canopy, thereby controlling soil loss and land degradation and improving soil health.

One of the benefits of soil/water conservation measures in the watershed was significant reduction (34.6%) in runoff and soil loss from the treated watershed area compared to the untreated area (Table 4.3). In case of untreated watershed area, 12% of rainfall was lost as runoff, whereas in treated area only 7.8% of rainfall was lost as runoff indicating 4% more rainwater was stored in soil which would have benefitted crop as well as part groundwater recharge. Data during 1999–2017 show soil loss of 2.75 t ha^{-1} in untreated area compared to 1.41 t ha^{-1} in treated watershed area recording 48.8% reduction in soil loss due to integrated watershed development in Adarsha watershed, Kothapally. Due to watershed development activities, Kothapally field retained on an average 1.34 t soil per ha per year which works out to be 25.5 t soil retention storing 0.4 t valuable organic carbon per ha in soil in 19 years since development contributing significantly to minimizing land degradation and sustainable crop yields. When considered at watershed level, in 19 years 11,840 t soil was retained in the watershed which contained 186 t of valuable soil organic carbon along with associated soil nutrients like N, P, K, Zn, B, Fe, S, Ca, Mn, Mg, etc., which are critical for sustainable crop yields.

Scientists from National Remote Sensing Agency (NRSA), Hyderabad and ICRISAT jointly, developed a remote sensing and GIS-based model for estimating soil loss from the small agricultural watersheds (Dwivedi et al. 2005). This model was used to assess the impact of watershed interventions on soil loss and land deg-

Table 4.3 Seasonal rainfall, runoff and soil loss from the sub-watershed in Adarsha watershed, Kothapally during 1999–2017

Year	Rainfall (mm)	Runoff (mm)		Soil loss (t ha^{-1})	
		Untreated	Treated	Untreated	Treated
1999	584	16	NR	NR	NR
2000	1161	118	65	4.17	1.46
2001	612	31	22	1.48	0.51
2002	464	13	Nil	0.18	Nil
2003	689	76	44	3.2	1.1
2004	667	126	39	3.53	0.53
2005	899	107	66	2.82	1.2
2006	715	110	75	2.47	1.56
2007	841	115	82	4.5	2.09
2008	1387	281	187	8.94	4.5
2009	710	130	83	2.30	1.90
2010	984	150	89	2.50	2.10
2011	574	40	26	2.10	1.10
2012	716	105	71	2.45	1.90
2013	775	98	60	3.06	1.67
2014	453	10	2	1.00	0.50
2015	491	50	3	0.90	0.10
2016	762	82	30	1.10	0.30
2017					
Mean	749	90.2	59.0	2.75	1.41

[a]*Untreated*, control with no development work, *treated* with improved soil/water/crop management, *NR* not recorded

radation. In this model, the digital elevation map was derived from panchromatic sensor (PAN) stereo data of Indian Remote Sensing Satellite IRS-1C and aerial photographs. The input parameters (soil erodibility, drainage density, length and degree of slope, surface cover, vegetation index, agricultural practices and flow routing) required for the model have been derived through visual interpretation of aerial photographs. The slope factor was derived from digital elevation model generated from aerial photographs and PAN stereo images. This remote sensing and GIS process-based model was used to assess the impact of watershed interventions on soil loss and land degradation. The data from Kothapally watershed from 2000 to 2007 was used to assess the impact of watershed interventions on soil loss, vegetative cover and area under waterlogging flooding especially during high-rainfall events. It estimated that during 1999–2007 about 3820 tons of soil loss was saved due to various watershed interventions. This has significantly contributed to improving soil health and reducing land degradation. Due to various watershed interventions, the vegetation index in watershed improved by 38% compared to start of the project (1999). The watershed interventions were also found effective in reducing downstream flooding and formation of new rills and gullies in the watershed areas.

4.5   Overall Impacts of Soil Health and Other Management Practices on Crop Productivity

The impacts of integrated improved watershed management practices were evaluated in on-farm crop yields that included integrated nutrient management along with crop and water management as important intervention (Table 4.4). With improved watershed technologies during 1999–2002, farmers obtained two- to threefold higher crop yields in case of major crops like maize and sorghum as compared with the base year during 1998. In the case of maize intercropped with pigeon pea, improved practices resulted in two- to fourfold increase in maize yield compared with farmers' traditional practices where the yields ranged between 0.7 and 1.8 t ha^{-1}. Of all the cropping systems studied in Adarsha watershed, maize/pigeon pea intercropping systems proved to be the most beneficial where farmers could gain about Rs 16,500 and Rs 19,500 from these two systems, respectively.

4.6   Water Quality

Unabated N-fertilizer use and N-fertilizer-based pollution due to leaching of nitrates into groundwater is an issue of concern globally. An assessment of holistic approach adopted in Adarsha watershed, Kothapally, during 1999–2016, showed a significant decrease in nitrate-N loss to 7.1 kg ha^{-1} in the treated watershed area compared to 13.5 kg ha^{-1} in untreated area. Adarsha watershed, Kothapally, is a site of learning that holistic and integrated interventions like conservation of soil/water resources along with soil-C building, INM and good practices in crop management are instrumental in enhancing N-use efficiency for more food production, while enhancing the water quality through reduced nitrate levels.

4.7   Creating Awareness and Capacity Building: Key for Success

Creating awareness and strengthening capacity of stakeholders is crucial in scaling out the impacts of soil management. Following the principle of "*seeing is believing*", the exposure visits of farmers to on-station watersheds at ICRISAT campus that improved management practices helped them understand and believe the unexploited potential in agriculture. Participatory approach was adopted to bring in the ownership by farmers. Participatory soil sampling and use of stress-tolerant pigeon pea cultivar were taken as an entry point activity because it involved majority stakeholders leading to tangible economic benefits as a result of soil health mapping-based fertilizer management. Farmer meetings and specialized training programmes on nutrient recommendations and fertilizer management, recycling wastes through

Table 4.4 Average yields with improved technologies in Kothapally watershed, 1999–2007

Crop/system	Before – 1998	1999– 2000	2000– 2001	2001– 2002	2002– 2003	2003– 2004	2004– 2005	2005– 2006	2006– 2007	Mean	CV %	SE±
Crop yield (kg ha^{-1})												
Improved system												
1. Sole maize	–	3250	3756	3300	3481	3920	3421	3920	3635	3585	16.15	75.35
2. Maize/pigeon pea intercrop system	–	5263	6483	5596	5652	6292	4989	6388	6165	5853	17.67	134.04
3. Sorghum/pigeon pea intercrop system	–	5010	6524	5826	–	5782	4795	5288	5308	5505	13.69	161.30
4. Sole sorghum	–	4358	4590	3574	2964	2745	3022	2864	2503	3327	23.90	141.38
Farmers' practice												
5. Sole maize	1500	1700	1601	1630	1661	1721	1951	2250	2151	1833	33.09	104.60
6. Sorghum/pigeon pea intercrop system	1983	2329	2174	2763	3188	3311	2998	3359	3118	2904	19.17	110.26
7. Hybrid cotton	–	2295	7047	6605	6487	6947	–	–	–	5876	28.48	286.97
8. BT cotton	–	–	–	–	–	–	–	6207	5590	5899		
CV%		24.9	30.1	10.0	8.0	15.6	20.5	11.5	11.0			
SE±		867	1497	383	323	752	748	529	475			

Source: Wani et al. (2006)

composting, biomass generation through *Gliricidia* and soil conservation measures built specialized skills amongst the farmers' community. Lead farmers especially played a key role in liaising between experts and farmers who generally follow other fellow farmers.

4.8   Summary and Important Findings

In rain-fed areas in real-world field situation, the soil depth varies based on the location of the fields on different topo-sequence and the soil fertility as well as water holding capacity differed a lot. Such situation calls for site-specific fertilizer management strategies rather than the crop or agro-ecoregion-based fertilizer recommendations. In order to meet the growing demand for food and nutrition security, available but untapped potential of dryland agriculture need to be harnessed. There are large and economically exploitable yield gaps in the drylands. These gaps can be easily abridged with current levels of technologies if holistic and integrated solutions adapted to local conditions are made available to farmers. Kothapally watershed is a typical example of such a pilot. In the watershed, it became evident that soil resources are badly deteriorated and can no longer be ignored to meet the challenges of increasing productivity and incomes on a sustainable basis. The focus was on addressing the issues of soil erosion and loss of soil fertility aggravated by uninformed decisions leading to mismanagement of fertilizers. An accurate diagnosis leading to need-based use of resources not only led to high productivity and profitability but also efficient and sustainable resource use that resulted in improving soil health over the years. It also demonstrated that state-of-the-art facilities are essential for diagnosing the nutrient deficiencies, or else it may just be a futile exercise. Policy support that focuses on conserving soil loss due to erosion by adopting on-farm and community-based interventions such as integrated watershed management strategy along with promoting balanced and integrated use of chemical fertilizers and recycling of on-farm wastes showed tangible economic benefits as well as conserved the natural resources in the watershed. Adarsha watershed, Kothapally, is an exemplar of pilot site of learning for soil health management for higher and sustained productivity and has helped in scaling up the learnings and strategies for adaptation to climate variability as well as climate change impacts. Soil management is a topic that has major implications on various ambitious sustainable development goals like no poverty, zero hunger, good health and well-being, clean water and sanitation, gender equality, decent work and economic growth, life on land, climate action and thereby needs major focus.

References

Bhattacharyya, R., Ghosh, B. N., Mishra, P. K., Mandal, B., Rao, C. S., Sarkar, D., Das, K., Anil, K. S., Lalitha, M., Hati, K. M., & Franzluebbers, A. J. (2015). Soil degradation in India: Challenges and potential solutions. *Sustainability, 7*, 3528–3570.

Chander, G., Wani, S. P., Sahrawat, K. L., Kamdi, P. J., Pal, C. K., Pal, D. K., & Mathur, T. P. (2013). Balanced and integrated nutrient management for enhanced and economic food production: Case study from rainfed semi-arid tropics in India. *Archives of Agronomy and Soil Science, 59*(12), 1643–1658.

Chander, G., Wani, S. P., Krishnappa, K., Sahrawat, K. L., Pardhasaradhi, G., & Jangawad, L. S. (2016). Soil mapping and variety based entry-point interventions for strengthening agriculture-based livelihoods – Exemplar case of 'Bhoochetana' in India. *Current Science, 110*(9), 1683–1691.

Chander, G., Wani, S. P., Gopalakrishnan, S., Mahapatra, A., Chaudhury, S., Pawar, C. S., Kaushal, M., & Rao, A. V. R. K. R. (2018). Microbial consortium culture and vermi composting technologies for recycling on-farm wastes and food production. *International Journal of Recycling of Organic Waste in Agriculture*. https://doi.org/10.1007/s40093-018-0195-9.

Dixit, S., Wani, S. P., Rego, T. J., & Pardhasaradhi, G. (2007). Knowledge-based entry point and innovative up-scaling strategy for watershed development projects. *Indian Journal of Dryland Agriculture and Development, 22*(1), 22–31.

Dwivedi, R. S., Sreenivas, K., & Pathak, P. (2005). *Quantitative estimation of soil loss from watershed* (Scientific report, 25 p). Hyderabad: National Remote Sensing Agency, Department of Space, Government of India., NRSA-RS&GIS-PR-707-I01.

ICRISAT (International Crops Research Institute for the Semi-Arid Tropics) (2002, September). *Improving management of natural resources for sustainable rainfed agriculture*. TA RETA 5812 completion report submitted to Asian Development Bank, The Philippines, 94 pp. (Limited distribution).

Joshi, P. K., Wani, S. P., Chopde, V. K., & Foster, J. (1996). Farmers' perception of land degradation: A case study. *Economic and Political Weekly, 31*(26, June 29), A-89–AA92.

Nagavallemma, K. P., Wani, S. P., Lacroix, S., Padmaja, V. V., Vineela, C., Babu Rao, M., & Sahrawat, K. L. (2006). Vermicomposting: Recycling wastes into valuable organic fertilizer. *SAT EJournal, 2*(1), 1–16.

Sahrawat, K. L., Rego, T. J., Wani, S. P., & Pardhasaradhi, G. (2008). Stretching soil sampling to watershed: Evaluation of soil-test parameters in a semi-arid tropical watershed. *Communications in Soil Science and Plant Analysis, 39*(2950–2960), 2008.

Shiferaw, B., Anupama, G. V., Nageswara Rao, G. D., & Wani, S. P. (2002). *Socioeconomic characterization and analysis of resource-use patterns in community watersheds in semi-arid India*. (Working Paper Series No. 12, p. 44). Andhra Pradesh, India: International Crops Research Institute for the Semi-Arid Tropics.

Sreedevi, T. K., Shiferaw, B., and Wani, S. P. (2004). *Adarsha watershed in Kothapally: Understanding the drivers of higher impact* (Global Theme on Agroecosystems Report no. 10, p. 24). Patancheru: International Crops Research Institute for the Semi-Arid Tropics (ICRISAT).

Wani, S. P., & Garg, K. K. (2015). Restoring lands and livelihoods in rain-fed areas through community watershed management. Living Land (pp. 97–100). Tudor Publisher, UK.

Wani, S. P., & Patil, M. (2018). Securing soils through people-centric watershed management for sustainable agricultural development. *A Better World, 4*, 28–32. Tudor Rose, U.K.

Wani, S. P., Sreedevi, T. K., Singh, H. P., & Pathak, P. (2002). *Farmer participatory integrated watershed management model: Adarsha Watershed, Kothapally, India – A success story* (22 pp). Patancheru: International Crops Research Institute for the Semi-Arid Tropics.

Wani, S. P., Pathak, P., Jangawad, L. S., Eswaran, H., & Singh, P. (2003a). Improved management of Vertisols in the semi-arid tropics for increased productivity and soil carbon sequestration. *Soil Use and Management, 19*, 217–222.

Wani, S. P., Singh, H. P., Sreedevi, T. K., Pathak, P. Rego, T. J., Shiferaw B., & Iyer, S. R. (2003b). *Farmer participatory integrated watershed management: Adarsha watershed, Kothapally, India.* (An innovative and upscalable approach INRM case study 7). *Journal of SAT Agricultural Research, 2*(1), 1–27

Wani, S. P., Singh, H. P., Sreedevi, T. K., Pathak, P., Rego, T. J., Shiferaw, B., & Iyer, S. R. (2006). Farmer-participatory integrated watershed management: Adarsha watershed, Kothapally India. An innovative and upscalable approach. *SAT E-Journal, 2*(1), 1–26.

Wani, S. P., Sahrawat, K. L., Sreeedevi, T. K., Pardhasaradhi, G., & Dixit, S. (2009). Knowledge-based entry point for enhancing community participation in integrated watershed management. In*: Proceedings of the comprehensive assessment of watershed programs in India, 25–27 July 2007* (pp. 53–68). Patancheru: International Crops Research Institute for the Semi-Arid Tropics. ISBN: 978-92-9066-526-7: CPE 167.

Wani, S. P., Chander, G., & Vineela, C. (2014). Vermicomposting: Recycling wastes into valuable manure for sustained crop intensification in the semi-arid tropics. In R. Chandra & K. P. Raverkar (Eds.), *Bioresources for sustainable plant nutrient management* (pp. 123–151). Delhi: Satish Serial Publishing House.

Wani, S. P., Chander, G., Bhattacharyya, T., & Patil, M. (2016). *Soil health mapping and direct benefit transfer of fertilizer subsidy* (Research report IDC-6). Patancheru: International Crops Research Institute for the Semi-Arid Tropics. Available at: http://oar.icrisat.org/9747/1/2016-088%20Res%20Rep%20IDC%206%20soil%20health%20mapping.pdf

Wani, S. P., Chander, G., & Pardhasaradhi, G. (2018). Soil amendments for sustainable intensification. In A. Rakshit, B. Sarkar, & P. Abhilashis (Eds.), *Soil amendments for sustainability: Challenges and perspectives.* Boca Raton: CRC Press/Taylor & Francis Group.

Chapter 5
Improved Water Balance and Ecosystem Services Through Integrated Watershed Development

Kaushal K. Garg, K. H. Anantha, S. P. Wani, Mukund D. Patil, and Rajesh Nune

Abstract Agricultural water management (AWM) interventions in Kothapally watershed enhanced provisional, regulating, and supporting ecosystem services. Kothapally watershed, which was in degraded stage before 1999, is transformed into highly productive stage through science-led natural resource management interventions. A number of AWM interventions such as field bunding, low-cost gully control structures and masonry check dams, etc., were built as per hydrological assessment and need of the community. Ridge to valley approach of rainwater harvesting addressed equity issue as farmers from upstream end benefited along with downstream users. A number of AWM interventions reduced surface runoff (30–60%) and soil loss (two- to fivefolds) and enhanced groundwater recharge (50–150%) and base flow. Water table increased from 2.5 to 6.0 m on an average after the AWM interventions. This change has translated into surplus irrigation water availability and crop intensification especially during post monsoonal season. Further, all such changes translated into better crop yield, higher cropping intensity, and higher crop production and net income over the years that resulted into building the resilience of the individuals and community to cope with droughts and impacts of climate change. This case study clearly indicates that large untapped potential exists in dryland areas which could be harnessed through science-led NRM interventions. Scaling up approach of these interventions through pilots at various locations in India, Thailand, Vietnam, and China demonstrated the potential for overcoming food and water scarcity sustainably and at the same time contributing to meet the sustainable development goals of zero hunger, water availability, and climate actions.

K. K. Garg (✉) · K. H. Anantha · M. D. Patil · R. Nune
ICRISAT Development Centre, Research Program Asia, International Crops Research
Institute for the Semi-Arid Tropics (ICRISAT), Hyderabad, Telangana, India
e-mail: k.garg@cgiar.org

S. P. Wani
Former Director, Research Program Asia and ICRISAT Development Centre, International
Crops Research Institute for the Semi-Arid Tropics (ICRISAT), Hyderabad, Telangana, India

© Springer Nature Switzerland AG 2020
S. P. Wani, K. V. Raju (eds.), *Community and Climate Resilience
in the Semi-Arid Tropics*, https://doi.org/10.1007/978-3-030-29918-7_5

Keywords Water balance · Groundwater recharge · Hydrologcial model · SWAT · Building system resilience

5.1 Introduction

Ensuring global food security and simultaneously maintaining ecosystem services is a challenging task for sustainable development. Crop intensification practices on one hand have increased "provisioning" ecosystem services, such as food, fiber, and timber production; however, it substantially impacted other ecosystem services negatively. The increased food production has come by depletion of finite and scarce water and land resources. For example, total cultivable land in India is 141 million ha with a cropping intensity of 135%. Groundwater and surface water sources irrigate about 41 and 21 million ha of agricultural lands (total 45%), respectively, and the rest (55%) of the cultivable area is rainfed.

The Green Revolution during the 1970s along with change in cropping pattern made a significant impact on groundwater use especially in dryland areas. The number of bore wells increased from less than one million during the 1960s to 20 million by 2009 in India (Dewandel et al. 2010). As a result, groundwater withdrawals escalated from less than 25 km^3 in the 1960s to 300 km^3 in 2008, which is several times higher than withdrawals of any other developed and developing country in the world (Shah 2009). During this development process, groundwater uses enhanced food production in the country, but in many of the Indian states/regions, there was a decline in groundwater sustainability (Garg and Wani 2013).

Presently, all the blue water is almost utilized/allocated in different sectors and there is not much scope for augmenting it. However, the large untapped potential in the rainfed areas holds a big hope for addressing food security in near future. This could be achieved through scientific land, water, and nutrient management interventions. Decentralized agricultural water management (AWM) interventions at field and community scale are not only helpful in strengthening provisional ecosystem services (providing food, fodder and timber, etc.) but also in regulating and supporting ecosystem services; those are essential for resilient and sustainable agro-ecosystem and also required for sustainability of provisional services (Singh et al. 2014).

Realizing these facts, a large-scale public investment has been made in India for implementing national scale program such as Integrated Watershed Management Program (IWMP) targeting for drought proofing, soil conservation, low-cost water harvesting and groundwater recharge in dryland regions since the 1970s onwards. Comprehensive assessment of watershed programs in India indicated that such interventions had benefit-cost (B:C) ratios of 2.0 or larger, enhanced ecosystem services (ECSs), and generated employment. However, 66% of the projects performed below average (Wani et al. 2008, 2012).

This chapter is largely focusing on the impact of various AWM interventions on water balance components and ecosystem services generated in Adarsha watershed, Kothapally. The local-scale interventions positively altered the microscale hydrology and enhanced crop intensification and higher productivity. These interventions also generated other essential ecosystem functions such as regulating and supporting services at landscape level which are described in details.

5.2 AWM Interventions at Community Watershed, Kothapally

ICRISAT and its consortium partners (national agencies, government line department, and NGO) described in chapter two of this volume started watershed development program in Kothapally from 1999 onwards, as a site of learning for various stakeholders such as farmers, development agencies, researchers, and policy makers. A range of agricultural water management initiatives have been adopted since it started, both at community and individual farm levels. The most common in situ interventions are contour and graded bunding, which reduce travel distance for runoff water, minimize the runoff velocity, and allow more water to percolate into the soil increasing soil moisture content (green water). This practice allows surface runoff accumulation along the bund and checks soil erosion. Check dams on the stream network reduced peak discharge and increased groundwater recharge. At the same time, these dams trapped sediments which protect the river ecosystems further downstream. The community made a protocol that water in the check dams will not be used directly for irrigation, and the stored water is allowed to recharge the groundwater aquifer by percolation. Instead, groundwater from open wells is used to irrigate crops (Garg et al. 2012). A large number of biophysical (topography, soil physical and chemical properties, land use), meteorological (rainfall, max temp, min temp, wind speed, relative humidity, solar radiation), hydrological (surface runoff and soil loss at watershed outlet by automatic gauging, groundwater table in 62 open wells), and socioeconomic (yield, income) data were collected starting from the initial phase of the project. Data collected from Kothapally watershed were analyzed using the SWAT model (Garg et al. 2012) and also the impact of AWM interventions on watershed hydrology and soil loss. After the model calibration, various land and water management scenarios were developed. Model inputs, calibration, and validation are described in detail in Garg et al. (2012) (Fig. 5.1).

5.3 Rainfall Amount and Its Variability

Rainfall is highly erratic, both in terms of total amount and its distribution over time. Mean annual rainfall at Kothapally is 820 mm with about 85% falling from June to October. Rainfall data show that a total of 450–1090 mm (average 741 mm) precipitation was received during the monsoon period (Jun–Oct) from years 1998 to 2017. Maximum rainfall intensity varied from 39 mm day^{-1} (in 2002) to 302 mm day^{-1} (in 2000), the latter representing an extreme event. Rainfall data showed that

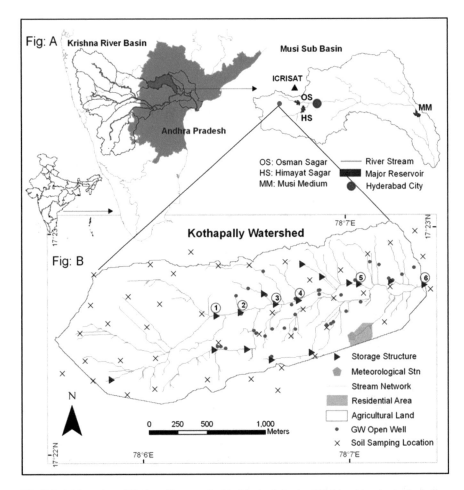

Fig. 5.1 (**a**) Location of Kothapally watershed in Musi sub-basin of Krishna river basin, including main reservoirs, ICRISAT, and Hyderabad City;. (**b**) Stream network, location of storage structures, open wells, meteorological station, soil sampling locations, and residential area in Kothapally watershed

dry spells longer than 5–7 days are very common and occur several times (three to eight times) per season, whereas 10–15 days or longer dry spell also may occur during the monsoon period.

Rainfall events are categorized into three broader classes (low, 0–25 mm day^{-1}; medium, 25–50 mm day^{-1}; and high, >50 mm day^{-1}) and shown between years 1998 and 2017 (Fig. 5.2a) along with trend lines (Fig. 5.2b). The data indicated that there is no significant change in rainfall pattern in low rainfall category, whereas medium-intensity events (25–50 mm day^{-1}) have increased significantly and high rainfall events (>50 mm) declined over the period. This has a large implication towards hydrological cycle. As high rainfall events are largely contributing to surface runoff, there could be significant reduction possible in runoff generation and blue water availability at downstream reservoirs in the region.

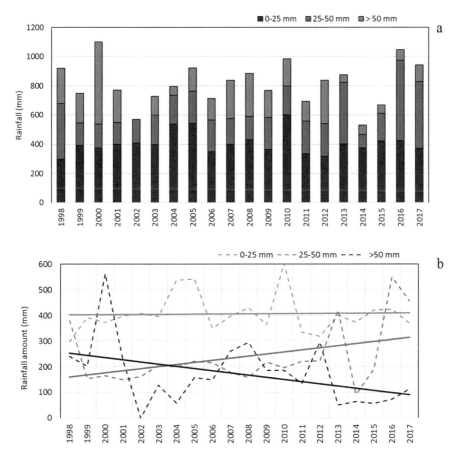

Fig. 5.2 (**a**) Rainfall category (low, medium, and high) and its temporal distribution over the years; (**b**) trend lines

5.4 Impact of AWM Interventions on Water Balance Components

Different soil and water conservation interventions significantly changed the water balance components in the Kothapally watershed. Results are presented for dry, normal, and wet years according to the IMD, Pune, India (http://www.imdpune.gov.in): rainfall less than 25% of the long-term average = dry; rainfall between −25% and +25% of the long-term average = normal; rainfall greater than 25% of long-term average = wet.

The water balance is affected by management interventions (Fig. 5.3). For the degraded state (no intervention), close to 64% of the rainfall was partitioned into ET, while just about 9% (70 mm) recharged the groundwater aquifer and 19%

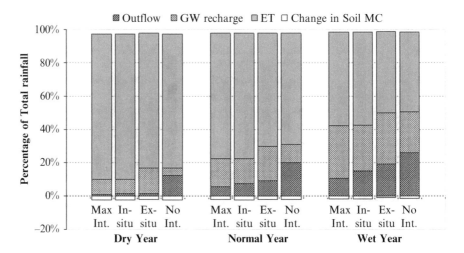

Fig. 5.3 Water balance for the four different water management scenarios for the first cropping season (from June to Dec) of different water years: dry, normal, and wet years. Max int.: in situ + check dams. In situ: in situ + no check dams. Ex situ: no in situ + check dams. No int.: no in situ + no check dams. *GW recharge* groundwater recharge. *ET* evapotranspiration. Change in SMC = change in soil moisture content

(151 mm) was lost from the watershed boundary as outflows during the cropping season.

When the watershed development program was in place (maximum intervention), the amount of water partitioned as ET had increased to around 576 mm, equivalent to 72% of annual average rainfall. Groundwater recharge was also higher (174 mm), while outflow from the watershed was less than 10% (70 mm or less than half of what it was before). Constructing check dams (ex situ) substantially increased groundwater recharge, while reducing outflows. In situ practices resulted in a higher ET, since more water was available as soil moisture in the fields, higher groundwater recharge, and lower outflow.

The water partitioning differs significantly between dry, normal, and wet years (Fig. 5.3). A large fraction of the total rainfall amount (85–90%) is converted into ET, while only a fraction generated outflow and groundwater recharge in dry years. On the other hand, only about 50–60% of the total rainfall was converted into ET during wet years. In the non-intervened watershed, outflow is small (<5–8% of the total rainfall) in dry years, but with interventions, outflow is almost negligible. During normal and wet years, outflow is reduced by 30–60% with interventions compared with the non-intervened state.

Groundwater recharge varies between 50 mm and 300 mm for dry and wet years, respectively (Fig. 5.4). Thus, the variation in groundwater recharge is larger between years than between treatments. During dry years, water management interventions became particularly important for groundwater recharge, which was more than twice as high for both ex situ and in situ interventions compared with the degraded state. Groundwater availability impacts the potential to grow a second, fully irrigated

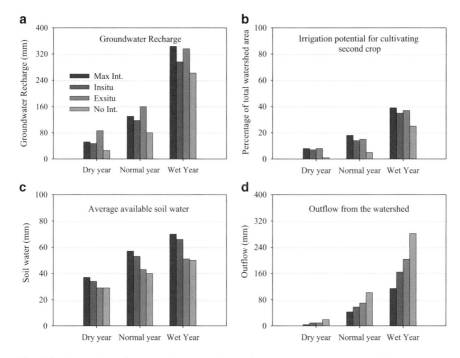

Fig. 5.4 Comparison of (**a**) groundwater recharge, (**b**) developed irrigation potential, (**c**) average available soil water during crop period, and (**d**) outflow amount in different land management scenarios during dry, normal, and wet years

crop during the dry season (Fig. 5.4b). Again, the variation between years is larger than between treatments. The irrigation potential is more than doubled with water management interventions during dry and normal years.

In situ water management resulted in higher soil moisture availability (Fig. 5.4c). This pattern was same for dry, normal, and wet years. Ex situ water management had a small impact on soil moisture availability for all years.

Outflow varies significantly between years and with AWM interventions (Fig. 5.4d). Outflow was more than ten times higher during wet years compared with dry years. With AWM interventions, outflow from the watershed was reduced by more than 50% compared to that in the non-intervened state. A linear relationship was found between rainfall and runoff from the watershed on annual basis (Fig. 5.5), but the magnitude varied with the AWM interventions on the field scale. The lowest outflow occurred with both check dams and in situ water management in place (max int.), while the non-interventions generated the highest outflow. Moreover, the result shows that runoff losses were smaller for in situ management compared to ex situ interventions, indicating that in situ management was more efficient in reducing runoff loss, as compared to check dams.

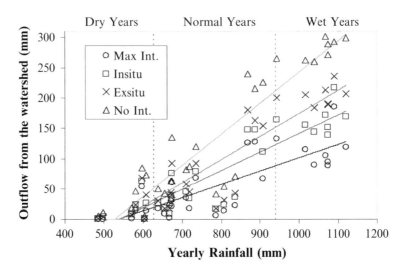

Fig. 5.5 Rainfall-runoff relationship for the four different water management scenarios. Results based on 31 years of simulation run from year 1978 to 2008. Max int.: in situ + check dams. In situ: In situ + no check dams. Ex situ: no in situ + check dams. No int.: no in situ + no check dams

5.5 Groundwater Dynamics and Resilience Building

Groundwater recharge in relation to cumulative rainfall presented for a selected normal year (2009) showed that over 300–400 mm of rainfall during the monsoon was required to cause a rise in water table (Fig. 5.6a). Data further showed that a large fraction of monsoonal rain was captured by soil layers initially and lost through evaporation and plant transpiration. After saturating the soil moisture profile, surplus water percolated down and recharged groundwater.

Groundwater recharge varied between years and with water management interventions (Fig. 5.6b). A direct linear relationship was found between rainfall and groundwater recharge both for no interventions and AWM interventions stages. With AWM interventions higher recharge was found especially in dry years (nearly double), but this difference in wet year was less significant. Further, grouping dry and normal years following a wet and dry year demonstrated carryover storage which was found significantly higher following a wet year, and this amount was further increased with AWM interventions in the watershed. AWM interventions helped in enhancing groundwater availability by recharging more water.

Water availability and crop yield have substantially improved after the watershed development program was implemented (Sreedevi et al. 2004; Garg et al. 2012; Karlberg et al. 2015). For instance, many of the wells that were not functioning due to deep groundwater have reverted into active well with good water yield. Farmers have switched over from cultivating cotton of traditional varieties and have started cultivating higher yielding cotton varieties (BT cotton) as irrigation water supply from the wells is now reliable. Those farmers who do not have wells for irrigation

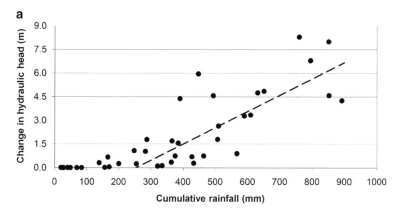

Fig. 5.6a Groundwater recharge vs. cumulative rainfall in Kothapally watershed

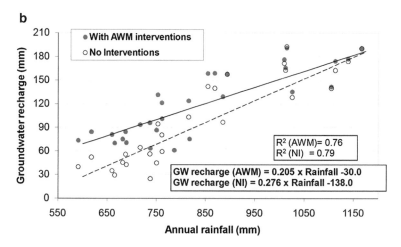

Fig. 5.6b Groundwater recharge in relation to monsoonal rainfall under no intervention (NI) and AWM intervention stage in Kothapally watershed

also started cultivating cotton as the soil moisture availability has improved after the in situ conservation practices which allowed crop to survive for another 2–3 months after monsoon without irrigation (Fig. 5.7).

A comparison of cotton yields for no intervention and after intervention scenarios is made among different years (dry, normal, and wet years). Results clearly show that nearly 50% farmers are getting cotton yield more than 2 t ha^{-1} after interventions compared to merely 30% before interventions in a normal year. Further the yield variation among these scenarios is shown spatially in Fig. 5.8.

Maize and paddy are now cultivated only in limited areas during monsoon and vegetables are grown in irrigated area during summer.

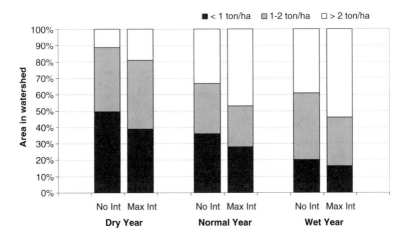

Fig. 5.7 Cotton crop yield in Kothapally watershed under no int. and max int. scenario

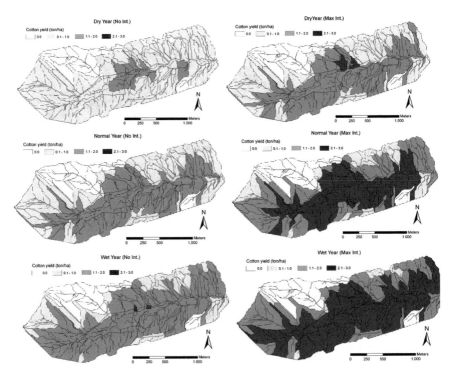

Fig. 5.8 Comparison of kharif crop yield between no int. (cotton) and max int. (cotton) scenarios during dry, normal, and wet years

Table 5.1 Irrigation potential for cultivating post monsoonal vegetable crops before and after watershed interventions during dry, normal, and wet years (per cent area)

Years	No int.	Max int.
Dry	3	8
Normal	7	23
Wet	28	55

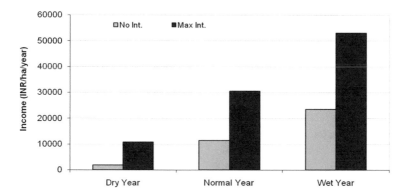

Fig. 5.9 Comparison of income generated from no int. and max int. scenarios. (Based on 30 years' average), cotton in monsoon season, and vegetable in post monsoon period

Irrigation potential for cultivating vegetable crop (post *kharif* season) is further described among different years in Table 5.1. There is significant groundwater availability for farmers to cultivate second season (vegetable) crop. For example, in a normal year, only 7% of farmers were able to cultivate vegetable with assured irrigation before watershed interventions, which has increased to more than three-folds, i.e., 23% fields, made a significant change in terms of enhanced land and water use efficiency and also total production.

Using *kharif* and *rabi* crop yields and by reducing the cost of cultivation, we estimated generated net income of the Kothapally farmers before and after watershed interventions among different years (dry, normal, and wet years) as shown in Figs. 5.9 and 5.10.

In the normal year, the net income generated before interventions was mere 11,000 Rs ha^{-1} year^{-1} which increased to more than 30,000 Rs ha^{-1} year^{-1}. Farmers in dry years were in deficit balance due to heavy risk of crop failure; with the interventions, it has converted to some net gain of nearly 10,000 Rs ha^{-1} year^{-1}. Net income in wet years was largely due to irrigation availability in post monsoon season and large area was converted into supplemental irrigation and resulted into more than 50,000 Rs ha^{-1} year^{-1} after the interventions compared to less than 25,000 Rs ha^{-1} year^{-1} before the watershed stage. Spatial variability of the income values from field to field and year to years are also shown in Fig. 5.10.

Furthermore, the livelihoods of the farmers in the Kothapally village have improved significantly with increased production and income. Crop yields are on

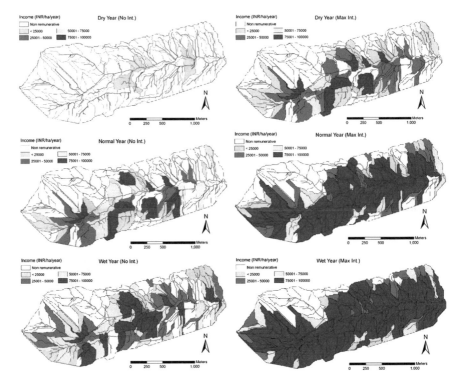

Fig. 5.10 Comparison of net income generated (INR/ha/year) among no int. and max int. scenarios. Both kharif and rabi crops are considered in this calculation: Net income = (crop yield x market price) - cost of cultivation. White color legend in given figures indicates that cost of cultivation is higher than the return achieved (cultivation is not remunerative)

the increase, and farmers are now able to save some of their farm income and reinvest in the business. Because of diversification of sources of income due to more off-farm activities, their resilience to external shocks has been improved. More specifically, the AWM interventions have reduced the inherent risks in agriculture in the semi-arid zone posed by high rainfall variability and frequent dry spells. With a more erratic precipitation under future climate change, AWM interventions in tropical agriculture are likely to be of greater importance.

The socioeconomic status has improved after introducing AWM in the Kothapally village (Sreedevi et al. 2004; Karlberg et al. 2015). Most of the farmers were solely dependent on agriculture in 1999 and before. Farmers were motivated to do other job activities and services along with cultivating crops, which together with improved yields led to a substantial change in their livelihood in recent years. Approximate one-fourth of households in Kothapally watershed generate additional income other than agriculture currently. The average household income in the Kothapally watershed is about 50% higher compared to adjoining villages without watershed interventions (Anantha and Wani 2016; Sreedevi et al. 2004). Moreover, despite high incidence of drought during 2002, the watershed interventions have

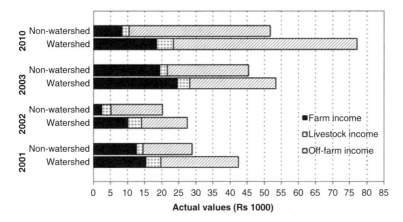

Fig. 5.11 Impact of watershed development approach on household income in Adarsha watershed (Source: Anantha and Wani 2016)

contributed to improved resilience of agricultural income. While drought-induced shocks reduced the average share of crop income in the non-watershed area from 44 to 12%, this share remained unchanged at about 36% in the watershed area. Livestock sector also contributed significantly to total household income in watershed villages even during drought situations (Sreedevi et al. 2004). Even after the withdrawal of watershed program, the contribution from crop activities is significant (>50%) in the watershed compared to non-watershed areas (Fig. 5.11). This signifies the importance of watershed interventions in dryland areas as a sustainable adaptive mechanism (Anantha and Wani 2016).

5.6 Groundwater Quality

An inappropriate use of nitrogenous (N) fertilizer has been interfering the natural N cycle. Several cases of nitrate contamination of groundwater as consequences of excessive use of N fertilizers have been reported in several parts of the world. Moreover, nitrate concentration as high as 450 mg L^{-1} is also observed in groundwater, but attributed to point sources, such as poultry farm, cattle shed, leakages from septic tank, etc. (Rao 2006). Increased concentration of nitrate in drinking water can cause several health hazards. Blue baby syndrome or methemoglobinemia is the major and fatal health issue caused by intake of nitrate-contaminated food and water (Fan and Steinberg 1996; Fewtrell 2004; Sadeq et al. 2008). Studies have also indicated that high nitrate levels in water may cause gastric cancer (Barrett et al. 1998; Sandor et al. 2001; van Loon et al. 1998). Apart from impact on human health, increased N concentration in water bodies may cause eutrophication and, thus, also has negative effects on aquatic life.

Generally, about 40–50% of plant N demand is satisfied by available N in soil and the rest of the N demand fulfilled through N fertilizers. The most common form of N fertilizer being used in developing countries is urea. The sequential chain reactions to transform the urea to the plant available N are hydrolysis of urea to form ammonium-N and nitrification of ammonium-N to form nitrate-N. The important loss mechanisms are volatilization of ammonium-N to gaseous ammonia, denitrification of nitrate-N to gaseous N, and transport of the dissolved N species via runoff and percolating water. High precipitation variability causes a considerable runoff event. If such event coincides with the fertilizer application, then the large amount of the applied N may wash out from the fields. This leads to an N-deficient situation inside the field, but N accumulation to the surface water body that is replenished by runoff water. These surface water bodies refilled by the runoff water from agricultural field become active recharge basins and, perhaps, source for both relatively faster nitrate leaching and denitrification than the fields.

In semi-arid regions efficient use of available water and rainfall is one of the key components for sustainable agriculture. Various in situ soil and water conservation techniques have been successfully implemented for optimizing the productivity of resources. One such technique is broad bed and furrow (BBF) system that has been successfully implemented on *Vertisols* in semi-arid region of India. The proper drainage through furrow and availability of stored moisture in broad bed take care of waterlogging due to excess rainfall and water scarcity due to extended dry spells. This also improves efficient rainwater usage. Runoff water may be captured by constructing water harvesting structures. In situ soil water regime improvised by conservation practices has great influence on transport and transformation processes.

The in situ and ex situ water management interventions have contributed to increased water availability leading to increase in cropping intensity, irrigation, and fertilizer usage. The increase in these factors is often reflected in the increase in N concentration in groundwater. But, as indicated earlier, rainwater harvesting interventions have improved groundwater level and the dilution of N species reduces the N level to safe limits. This regulatory ecosystem service of water management interventions was studied by monitoring groundwater quality of selected open and bore wells in Kothapally village. The open wells, often, receive return flow from field while irrigation and sometimes runoff from cropland during heavy rainfall event.

The average concentration of nitrate-N in open and bore wells for 4 years is presented in Fig. 5.12. The nitrate levels are crossing the permissible limit of 10 mg L^{-1} for nitrate-N concentration, which is a more concerning issue and may be attributed to recent increase in cropping intensity and the fact that all of these wells are irrigation wells in the crop lands. Moreover, N concentration for both open and bore well was similar. This indicated that the open wells were also acting as prime zone for groundwater recharge. The fluctuation in N concentration in the bore wells may be attributed to recharge of groundwater by N loaded water and by fresh rainwater with diluted N loading. In case of open wells, the seepage flow from vadose zone of croplands was the major source of water.

It was understood that the N fertilizers applied in the croplands served as major source for groundwater contamination, but a coarse frequency of monitoring

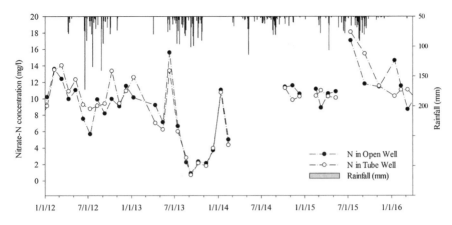

Fig. 5.12 Average nitrate-nitrogen concentration in open and bore well between January 2012 and March 2016

(monthly) could not capture the N concentration peaks with respect to fertilizer application. However, fluctuation in the N concentration in groundwater was following the rainfall trend, for example, the reduction in N concentration observed during July 2013, July 2014, and August 2015 (Fig. 5.12) was preceded by high or continuous rainfall events. Similarly, increase in N concentration during June 2013 and January 2014 might be due to irrigation water as open wells were receiving return flow of N loaded water from vadose zone of croplands.

The impact of rainwater harvesting on groundwater quality was seen in two ways, the one being improving the groundwater quality attributed to dilution of N concentration as the large amount of groundwater recharge takes place during rainy season and higher N loading per unit area as the increased water availability has increased the cropping intensity. Thus, better integration of nutrient and irrigation management practices will further improve the role of rainwater harvesting interventions in regulating the groundwater quality.

References

Anantha, K. H., & Wani, S. P. (2016). Evaluation of cropping activities in the Adarsha watershed project, southern India. *Food Security, 8*(5), 885–897.

Barrett, J. H., Parslow, R. C., McKinney, P. A., Law, G. R., & Forman, D. (1998). Nitrate in drinking water and the incidence of gastric, esophageal, and brain cancer in Yorkshire, England. *Cancer Causes & Control, 9*, 153–159.

Dewandel, B., Perrin, J., Ahmed, S., Aulong, S., Hrkal, Z., Lachassagne, P., Samad, M., & Massuel, S. (2010). Development of a tool for managing groundwater resources in semiarid hard rock regions: Application to a rural watershed in South India. *Hydrological Processes, 24*, 2784–2797.

Fan, A. M., & Steinberg, V. E. (1996). Health implications of nitrate and nitrite in drinking water: An update on methemoglobinemia occurrence and reproductive and developmental toxicity. *Regulatory Toxicology and Pharmacology, 23*, 35–43.

Fewtrell, L. (2004). Drinking-water nitrate, methemoglobinemia, and global burden of disease: A discussion. *Environmental Health Perspectives, 112*, 1371–1374. https://doi.org/10.1289/Ehp.7216.

Garg, K. K., & Wani, S. P. (2013). Opportunities to build groundwater resilience in the semi-arid tropics. *Groundwater, 51*(5), 679–691.

Garg, K. K., Karlberg, L., Barron, J., Wani, S. P., & Rockstrom, J. (2012). Assessing impacts of agricultural water interventions in the Kothapally watershed, Southern India. *Hydrological Processes, 26*(3), 387–404.

Karlberg, L., Garg, K. K., Barron, J., & Wani, S. P. (2015). Impacts of agricultural water interventions on farm income: An example from the Kothapally watershed, India. *Agricultural Systems, 136*, 30–38.

Rao, N. S. (2006). Nitrate pollution and its distribution in the groundwater of Srikakulam district, Andhra Pradesh, India. *Environmental Geology, 51*, 631–645. https://doi.org/10.1007/s00254-006-0358-2.

Sadeq, M., Moe, C. L., Attarassi, B., Cherkaoui, I., ElAouad, R., & Idrissi, L. (2008). Drinking water nitrate and prevalence of methemoglobinemia among infants and children aged 1–7 years in Moroccan areas. *International Journal of Hygiene and Environmental Health, 211*, 546–554. https://doi.org/10.1016/j.ijheh.2007.09.009.

Sandor, J., Kiss, I., Farkas, O., & Ember, I. (2001). Association between gastric cancer mortality and nitrate content of drinking water: Ecological study on small area inequalities. *European Journal of Epidemiology, 17*, 443–447.

Shah, T. (2009). Climate change and groundwater: India's opportunities for mitigation and adaptation. *Environmental Resources Letters, 4*, 035005.

Singh, R., Garg, K. K., Wani, S. P., Tewari, R. K., & Dhyani, S. K. (2014). Impact of water management interventions on hydrology and ecosystem services in Garhkundar-Dabar watershed of Bundelkhand region, Central India. *Journal of Hydrology, 509*, 132–149.

Sreedevi, T. K., Shiferaw, B., & Wani, S. P. (2004). Adarsha watershed in Kothapally, understanding the drivers of higher impact. In *Global theme on agroecosystems report no. 10*. Andhra Pradesh, India: International Crops Research Institute for the SemiArid Tropics.

Van Loon, A. J. M., Botterweck, A. A. M., Goldbohm, R. A., Brants, H. A. M., van Klaveren, J. D., & van den Brandt, P. A. (1998). Intake of nitrate and nitrite and the risk of gastric cancer: A prospective cohort study. *British Journal of Cancer, 78*, 129–135.

Chapter 6
Improved Livelihoods Through Sustainable and Diversified Cropping Systems

K. Srinivas, Gajanan L. Sawargaonkar, A. V. R. Kesava Rao, and S. P. Wani

Abstract Climate change presents an additional challenge for sustainable food production in the developing world. It is necessary to enhance present yield levels. Deriving and popularizing suitable cropping systems is critical. Integrated watershed management (IWM) implemented in Adarsha watershed, Kothapally, is a fine example of overcoming negative impacts of climate change. About 250 rainwater harvesting structures were constructed. About 27% of the rainfall contributed to groundwater recharge and risen the groundwater levels by 2–3 m. Due to increased water availability, farmers were able to diversify crops and grown two/three crops. In the post-rainy season, vegetables, rice, sorghum and chickpea were grown. Increased soil organic carbon (SOC) stocks were observed due to inclusion of legumes in cropping systems. According to the Intergovernmental Panel on Climate Change (IPCC), climate change affects crop production more negatively than positive, and countries like India are highly vulnerable. As per the general circulation model (GCM) CESM1-CAM5, RCP8.5, for Kothapally area, maximum and minimum temperatures during monsoon (Jun–Sep) are to be increased by 1.0 and 0.9 °C, respectively, by 2030. At the same time, rainfall in June and July together is expected to decrease by 65 mm, and August and September together are projected to have increased rainfall of about 50 mm. Sustainability of crops like cotton, maize, sorghum and pigeonpea was studied using crop simulation models. Impacts of climate change on productivity were estimated and suitable adaptation strategies were derived.

Keywords Cropping system · diversification · climate change · participatory evaluation

K. Srinivas (✉) · G. L. Sawargaonkar · A. V. R. Kesava Rao
ICRISAT Development Centre, Research Program Asia, International Crops Research Institute for the Semi-Arid Tropics (ICRISAT), Hyderabad, Telangana, India
e-mail: k.srinivas@cgiar.org

S. P. Wani
Former Director, Research Program Asia and ICRISAT Development Centre, International Crops Research Institute for the Semi-Arid Tropics (ICRISAT), Hyderabad, Telangana, India

© Springer Nature Switzerland AG 2020
S. P. Wani, K. V. Raju (eds.), *Community and Climate Resilience in the Semi-Arid Tropics*, https://doi.org/10.1007/978-3-030-29918-7_6

6.1 Introduction

Now the climate change is evident and presents new challenges to the world's growing population for food security (Ericksen et al. 2009). The world population is projected to reach 9.8 billion by 2050 and 11.2 billion by 2100 (United Nations 2015). Indian population is expected to reach 1.7 billion by 2050. Agriculture is the main source of livelihood for almost 56% of the country's total population, and most of the farming community are smallholders. Smallholder farming systems in arid and semi-arid areas are more vulnerable to climate change and likely to be adversely affected in these regions (Naab et al. 2012; Descheemaeker et al. 2016). Agricultural growth also has a direct impact on poverty eradication and is an important factor in employment generation. To improve the resilience of these farming systems, context-specific information is needed for effective decision-making and for the selection and implementation of strategies towards climate-smart agriculture (Lipper et al. 2014). Global temperature is expected to rise up to 5.5 °C by the end of the century under different scenarios (Fig. 6.1). For the major crops (wheat, rice and maize) in tropical and temperate regions, climate change without adaptation is projected to negatively impact production for local temperature increases of 2 °C or more above late-twentieth-century levels (IPCC 2014).

Adarsha watershed is located in Kothapally village (longitude 78°5′ to 78°8′ E and latitude 17°20′ to 17°24' N) in Ranga Reddy District of erstwhile undivided Andhra Pradesh, (now in Ranga Reddy District of Telangana). This watershed is a typical semi-arid area with annual rainfall of 826 mm with 624 mm during south-west monsoon season. This watershed spreads in an area of 465 ha of which 430 ha is cultivated and the remaining area is wasteland and settlement. Soils in Kothapally are predominantly Vertisols and associated soils (90%). The soil depth ranges from 30 to 90 cm and has medium to low water-holding capacities. The average landholding per household is 1.4 ha (Shiferaw et al. 2002). Cotton, maize, sorghum and pigeonpea are major annual crops grown in this watershed. The productivity levels, water use efficiency and groundwater levels were very low in this village.

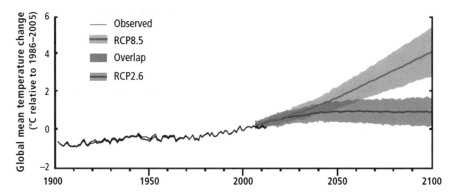

Fig. 6.1 Projected increase of global mean temperature (Source: IPCC)

ICRISAT-led consortium with national partners with farmers' centric approach has taken up the watershed development activities in this village in partnership with Drought Prone Area Programme (DPAP) of the state government. This consortium has implemented several water conservation structures like 14 water storage structures (one earthen and 13 masonry), 97 gully control structures, 60 mini percolation pits, 1 gabion structure and field bunds. In situ water conservation techniques like broad-bed and furrow (BBF) landform and contour planting were implemented. *Gliricidia* plants were grown on field bunds to strengthen bunds, conserve rainwater and supply nitrogen (N)-rich organic matter for in situ application to crops (Wani et al. 2002). Soil test-based micro- and macronutrients were applied. These activities resulted in increased crop yields in the range of 13–40%. Groundwater levels were dramatically improved which led to cultivation of crops during *rabi* (post-rainy) and summer seasons.

Climate change poses a great threat to agriculture and many studies have projected a decrease in crop yields. This study is undertaken to assess the impact of climate change on cotton, maize, sorghum and pigeonpea at Adarsha watershed, Kothapally. The future (2030 and 2050) projected rainfall and temperature data from a general circulation model (GCM) CESM1-CAM5 under the climate change scenarios RCP4.5 and RCP8.5 have been taken up. Crop simulation models Decision Support System for Agrotechnology Transfer (DSSAT) and Agricultural Production Systems sIMulator (APSIM) were used to assess the impact of projected climate change on crop yield and water balance. Temperature increases in the range of 0.7–3.5 °C, rainfall projected to be reduced during rainy season and increased during post-rainy season.

6.2   Climatic Features of Kothapally

Adarsha watershed at Kothapally is located at 17.375 °N latitude, 78. 119 °E longitude, and about 612 m above mean sea level in the Shankarpally Mandal, Ranga Reddy District of Telangana, India. Kothapally falls under hot moist semi-arid agroecological subregion (AESR) with deep loamy and clayey mixed red and black soils. Available water capacity of these soils is medium to very high (100–300 mm).

Daily rainfall data of Kothapally was collected for 20 years (1998–2017) and analysed. Average annual rainfall for Kothapally is about 826 mm. Normal monthly rainfall characters of Kothapally are shown in Table 6.1. Normal date of onset of

Table 6.1 Normal rainfall characters of Kothapally

Rainfall character	Jan	Feb	Mar	Apr	May	Jun	Jul	Aug	Sep	Oct	Nov	Dec
Average (mm)	4	6	20	28	44	112	150	206	156	88	10	2
Standard deviation (mm)	7	16	35	30	54	59	83	104	84	76	12	3
Coefficient of variation (%)	218	252	172	108	125	54	56	51	54	86	125	243
Rainy days	0	0	2	2	3	7	10	11	9	5	1	0

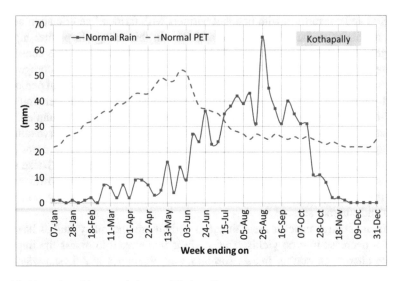

Fig. 6.2 Normal weekly water balance of Kothapally

southwest monsoon over Kothapally is around 5 June and the monsoon withdraws by the first week of November. Monthly rainfall distribution at Kothapally indicates that August is the rainiest month of the year with an average rainfall of about 206 mm followed by September with about 156 mm of rainfall. December to February is winter period which is dry having a low rainfall of about 12 mm and with high coefficient of variation of rainfall. During the 4 months June to September (southwest monsoon period), rainfall variability is less but still above 50 per cent. Rainy day is defined as a day which receives above 2.4 mm, and it is seen that there are only 50 rainy days at Kothapally, of which 30 rainy days are present during July to September, indicating the prime rain-fed crop-growing period.

Climatic water balance was worked out based on weekly rainfall and potential evapotranspiration (PET) and is shown in Fig. 6.2. Annual rainfall is about 818 mm, while the annual PET is about 1655 mm indicating that atmospheric and crop water requirements are very high compared to rainfall; only for about 14 weeks, rainfall exceeds PET starting from the second half of July. Second half of August in general receives high rainfall, and this is the time for water harvesting for use in the end period of *kharif* (rainy season) crops and for *rabi* crops.

6.3   Cropping System Management

Developing countries such as China, India, Brazil, Mexico, Indonesia, Vietnam, Pakistan and Sri Lanka face the great challenge in terms of climate-resilient technologies for food production. Although, fertility rates have dropped substantially in these countries, populations are continuing to grow rapidly in most of them as a consequence of demographic momentum. Moreover, all are experiencing increasing water demand for food production due to decreasing water availability along with increasing per capita demand for food as well as changing food habits with increasing incomes resulting in shift towards animal-based food products such as meat and milk. Therefore, balancing productivity, profitability and environmental health is a key challenge for today's agriculture for ensuring long-term sustainability (Foley et al. 2011). However, most crop production systems in the world are characterized by low species and management diversity, high use of fossil energy and agrichemicals and large negative impacts on the environment. Therefore, there is an urgent need to focus our attention towards the development of crop production systems with improved resource use efficiencies and more benign effects on the environment (Foley et al. 2011; Tilman et al. 2002) while meeting the goal of food and nutrition security. Cropping system design provides an excellent framework for developing and applying integrated approaches to management because it allows for new and creative ways of meeting the challenge of sustaining the agricultural productivity.

6.3.1 An Integrated Cropping System Management Approach at Production System Level

The innovative integrated approach was adopted in cropping system management in Kothapally through piloting science-based proven technologies to converge all the productivity enhancement interventions at production system level funnelled as a holistic complete package in order to unlock the potential of rain-fed agriculture. The adopted holistic approach provides a cost-effective and eco-friendly agrotechnological package to improve long-term rural productivity, profitability and employment on long-term basis in rural areas (Wani et al. 2003).

Looking at the changing complex scenario, in Kothapally, before the start of the IWM programme, a huge yield gap existed at farmers' field due to lack of outreach of the technologies developed by the agricultural universities, ICAR institutes and international centres. Secondly, natural resource management is like jazz as the stakeholders change their aspirations in context to exogenous factors which have unpredicted influences on farming. By considering all these challenges, it was revealed that the approach to promote the integrated genetic and natural resource management (IGNRM) interventions must be based upon continuous dialogue with the farmers (Holling and Meffe 1996; Wani et al. 2003, 2006, 2008) and deliberation among stakeholders (Hagmann et al. 1999) and needs to engage all the stakeholders in positive action to develop appropriate solutions together with resource users (Hagmann et al. 2002).

In this context, this watershed programme had the plan to make a difference in the lives of the farmers of Kothapally through the use of science-based technological approach along with sustainable natural resource management through participatory consortium approach. The programme has adopted an integrated approach in bringing together all the site-specific improved technologies to scale up the results of on-station as well as on-farm results obtained through demonstrations carried out in different states of the country. The interventions were basically having focus on identification and scaling up of best-bet options (soil, crop and water management) including improved cultivars to enhance the stagnated productivity of the selected crops in the entire village besides building capacity of the stakeholders (farmers and consortium partners) in the sustainable management of natural resources and enhancing productivity in dryland areas. The strategy to reach large number of farmers for adopting the developed integrated approach is achieved through using the concept of lead farmers as trainers to train large number of farmers. Therefore, in this chapter, all the interventions integrated and introduced at production system level in Kothapally were addressed.

6.3.2   Cropping System Management Through Farmers' Participatory Varietal Evaluation (FPVE)

The participatory varietal evaluation programme works towards increasing farm productivity by facilitating the delivery of high-yielding, profitable cultivars that are well adapted to a wide range of soil types, environment and farming systems. This is achieved by providing accredited, unbiased information to farmers on better adapted new crops and better cultivars, at the earliest opportunity. In Kothapally, farmers were given the choice to choose improved cultivars of preferred dryland crops from the list of cultivars provided to farmers' groups. Released improved cultivars and proprietary hybrids of crops were evaluated in this programme with an objective to select cultivars having suitable traits for better adaptation to biotic and abiotic stresses to enhance or sustain productivity and further scale up the spread of these cultivars to surrounding areas. Each demonstration was laid out approximately on half to one acre of farmers' field. Best-bet management includes application of 70 kg DAP, 100 kg urea, 5 kg borax, 50 kg zinc sulphate and 200 kg gypsum ha^{-1} for cereal crops and for legumes; a reduction in urea application from 100 kg to 40 kg ha^{-1} was done.

Increasing crop productivity is a common objective of all the watershed programmes, and enhanced crop productivity is achieved after the implementation of soil and water conservation practices, along with appropriate crop and nutrient management. For example, the implementation of improved crop management technology in the benchmark watersheds of Andhra Pradesh increased the maize yields 2.5 times (Table 6.2) and sorghum yields threefold [Wani et al. 2006]; thus implementing best-bet practices resulted in significant yield advantages, with varying crops from 63% to 197%. The similar crop responses were also recorded by Sreedevi and Wani (2009) and Wani et al. (2006) across different watersheds, and the increases ranged in sorghum from 35% to 270%, in maize from 30% to 174%, in pearl millet

Table 6.2 Average crop yields (kg/ha with equivalent of maize crop with different cropping systems at Adarsha watershed, Kothapally, (1999–2008)

Cropping systems	Yield (Kg/Ha) Before 1998	1999–2000	2000–2001	2001–2002	2002–2003	2003–2004	2004–2005	2005–2006	2006–2007	2007–2008	2008–2009	Mean	Cv%	SE
Improved systems														
Sole maize	–	3250	3760	3300	3480	3920	3420	3920	3630	4680	4810	3820	17.8	80
Maize/pigeonpea (intercrop system)	–	5260	6480	3570	5650	6290	4990	6390	6170	6120	6680	5960	16.7	116
Sorghum/pigeonpea (intercrop system)	–	5010	6520	5830	–	5780	4790	5290	5310	–	–	5500	13.4	154
Sole sorghum	–	4360	4590	3570	2960	2740	3020	2860	2500	–	–	3330	23.9	141
Farmer practice														
Sole maize	1500	1700	1600	1600	1800	1950	2250	2250	2150	–	–	1890	17.2	53
Sorghum/pigeonpea (intercrop system)	1980	2330	2170	2750	3190	3000	3360	3360	3120	–	–	2900	19.2	110
Hybrid cotton	–	2295	7050	6600	6490	6950	–	–	–	–	–	5880	37	511
BT cotton	–	–	–	–	–	–	–	6210	5590	7310	9380			315
Mean		3477	4970	3833	4010	4814	3651	4584	4320	6268	7390	–	–	–
CV%		11.9	31.4	10.7	8	14.5	20.3	10.8	12.2	16.7	16.2	–	–	–
SE		415	1559	410	323	698	742	495	525	1049	1201	–	–	–

The farmers' practice on sorghum/pigeonpea intercrop system; improved pigeonpea variety ICPL 87119 was grown along with local sorghum variety (*Pacha jonna*) from 2001 onwards. The old variety, which was highly susceptible for *Fusarium* wilt, was discontinued

from 72% to 242%, in groundnut from 28% to 179% in sole pigeonpea from 97% to 204% and in intercropped pigeonpea from 40% to 110%. A reduction in nitrogen fertilizer (90–120 kg urea per ha) by 38% increased maize yield by 18%.

6.4 Crop Diversification

Crop diversification in India is generally viewed as a shift from traditionally grown less remunerative crops to more remunerative crops. Potential benefits of diversifying cropping systems through efficient crop rotations as a means of increasing crop productivity and incomes while simultaneously enhancing other desirable agroecosystem processes (Liebman et al. 2001; Bennett et al. 2012; Karlen et al. 1994) were considered. Crop diversification involves incorporating potential substitute crop in the existing cropping system in order to maintain soil health, bring in stability in production system and reduce risk and insect pest incidence. Crop rotation is intended to give a wider choice in the production of a variety of remunerative crops in a given area so as to expand production-related activities on various crops and thereby ensure more profitability. Similarly, the aim in diversifying the cropping system, particularly in rain-fed ecology, is to reduce the risk factor of crop failures due to drought or less rains or market drop-down.

Analysis of prevalent cropping systems of Kothapally, their area and previous history before the watershed management intervention provided insight into the way the watershed management approach has benefited farmers. Kothapally was predominantly a cotton-growing area prior to project implementation. The area under cotton was 200 ha in 1998, whereas some area was also under maize, chickpea, sorghum, pigeonpea, vegetables and rice. After 4 years of activities in Adarsha watershed, the area under cotton cultivation decreased from 200 to 80 ha (60% decline) with simultaneous increases in maize and pigeonpea. The area under maize and pigeonpea increased more than threefolds from 60 to 200 ha and 50 to 180 ha, respectively, within 4 years. The area under chickpea also increased twofolds during the same period (Table 6.3) (Sreedevi et al. 2004).

Table 6.3 Area (ha) under various crops in Adarsha watershed, Kothapally (Sreedevi et al. 2004)

Crop	1998[a]	1999	2000	2001	2002[b]
Maize	60	80	150	180	200
Sorghum	30	40	55	65	70
Pigeonpea	50	60	120	180	180
Chickpea	45	50	60	60	100
Vegetables	40	45	60	60	100
Cotton	200	190	120	100	80
Rice	40	45	60	60	60

[a]Before watershed management activities began
[b]After 4 years of watershed management activities

Analysis of the data showed that average net returns per hectare for dry land cereals and pulses were significantly higher within the watershed. For cereals, the returns to family labour and land (net income) were 45% higher even with irrigation, while the net returns on rain-fed cereal crops were more than doubled (Table 6.4).Similarly, for pulse crops, per hectare net returns within the watershed were about twice as large as that outside the watershed. This was mainly because the integrated watershed development approach included improved cultivars of cereals (e.g. sorghum) and pulses (e.g. chickpea and pigeonpea) developed by ICRISAT along with improved management of water and soil fertility. Adoption of the improved cultivars has not only increased crop yields but also enhanced the economic profitability of other soil and water conservation investments, which may otherwise be economically unattractive to farmers.

In addition to the impacts on the net productivity of land, net incomes from crop production activities among the households within and outside the watershed were compared. The results are quite striking. Average household net income (without excluding family labour and owned land costs) from crop production activities within and outside the watershed was Rs 15,400 and Rs 12,700, respectively. The respective per capita income was Rs 3,400 and 1,900. Accounting for the cost of family labour, the average crop income within the watershed was Rs 12,700 compared to Rs 9,500 for the non-watershed villages (Fig. 6.3). Based on the baseline data from a random sample of 54 households, average net crop incomes (accounting for the cost of family labour) within the watershed in 1998, before the project started in the village, were computed. Average net crop income (in 2001 prices) in 1998 was about Rs 6,200 despite the high rainfall recorded in the village during that year (1084 mm vs 676 mm in2001). This shows that the average crop net income has doubled since 1998 (Sreedevi et al. 2004).

The technological change brought through availability of improved varieties, soil fertility and pest management practices and the increased availability of water has made substantial impacts on the livelihoods of the people in the village. Supplementary irrigation and new employment opportunities have also contributed to diversification of income and reduced vulnerability to drought and other stresses.

Table 6.4 Net income from crop production activities (Rs ha^{-1}) (Sreedevi et al. 2004)

	Within the watershed		Outside the watershed	
Crop irrigation	With irrigation	Without irrigation	With irrigation	Without irrigation
Cereals	11,170	6040	7690	2900
Pulses	8860	3810	4, 080	1920
Cotton	17,830	12,150	17,470	12,030
Vegetables	17,170	7480	11,980	6450
All crops	12,720	5880	14,810	3820

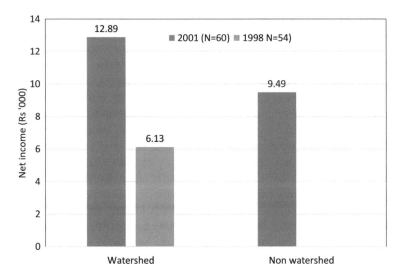

Fig. 6.3 Average household net income from crop production (2001 prices) in Adarsha watershed and surrounding villages (note: total variable cost includes family labour) (Sreedevi et al. 2004)

6.5 Soil and Water Conservation Methods

It is recognized that water shortage-related plant stress is the primary constraint to crop production and productivity in the rain-fed systems in the SA, and consequently the importance of water shortage has globally been rightly emphasized (Wani et al. 2002, 2003; Pathak et al. 2009). In rain-fed agriculture, accelerated demand for rainwater can be met through the efficient rainwater conservation and management. For this both in situ and ex situ rainwater management play crucial roles in increasing and sustaining the crop productivity. The comprehensive assessment of water management in agriculture (Comprehensive Assessment of Water Management of Agriculture 2007) describes a large untapped potential for upgrading rain-fed agriculture and calls for increased water investments in the sector. Based on experiences from the various watershed programmes and research station works in India, the soil and water conservation practices were identified and the information was used to determine the appropriate in situ and ex situ soil and water conservation practices in Kothapally.

The efforts were concentrated on different low-cost rainwater harvesting and in situ soil and water conservation measures which can be practically implemented and adopted by the farmers. Field-based soil and water conservation measures are essential for in situ conservation of soil and water. The main aim of these practices is to reduce or prevent either water erosion or wind erosion, while achieving the desired moisture for sustainable production (Table 6.5).

Table 6.5 Economics of production of different crops with improved technology in Adarsha watershed, Kothapally, during 1999–2000 (Sreedevi et al. 2004)

Cropping system	Total productivity (kg ha^{-1})	Cost of production (Rs ha^{-1})	Total income (Rs ha^{-1})	Profit (Rs ha^{-1})	Benefit-cost ratio
Maize/pigeonpea (improved)	3351	6203	22,709	16,506	2.67
Sorghum/ pigeonpea (improved)	2285	5953	17,384	11,431	1.92
Cotton (traditional)	980	15,873	24,389	8516	0.54
Sorghum/ pigeonpea (traditional)	1139	4608	11,137	6529	1.42
Maize-chickpea (improved)	4319	7317	26,774	19,457	2.66
Chickpea (improved)	840	4886	17,292	12,406	2.54
Sole maize (improved)	3150	4578	13,532	8954	1.96
Sorghum (traditional)	975	3385	6997	3612	1.07
Sole sorghum (improved)	2800	4352	15,084	10,732	2.47
Maize (traditional)	1600	3599	7281	3682	1.02
Mung bean (traditional)	600	4700	9000	4300	0.91
Chickpea (traditional)	–	4260	11,600	7340	1.72
Sole pigeonpea (improved)	1090	4890	17,120	12,230	1.35

6.5.1 Broad-Bed and Furrow and Related Systems

Soils with high clay content, viz. Vertisols, are often prone to waterlogging, thereby resulting into crop failure. Out of 30 districts of neighbouring Karnataka, around 12–13 districts are having Vertisols as predominant soil type and often face the waterlogging problem. An in situ soil and water conservation technology called "broad-bed and furrow system" has been promoted that can protect soil from erosion throughout the season and helps in proper drainage. This raised land configuration (BBF) system has been found to satisfactorily attain these goals. The BBF system consists of a relatively raised flat bed or ridge approximately 95 cm wide and shallow furrow about 55 cm wide and 15 cm deep (Fig. 6.4). The system is laid out on a grade of 0.4–0.8% for optimum performance. This BBF system is most effectively implemented in several operations or passes (Kampen 1982).

BBF formation with tropicultor   Groundnut crop on BBF

Fig. 6.4 The broad-bed and furrow system at Raichur, Karnataka, India

Table 6.6 Nutrient budgeting studies in farmers' fields, Adarsha watershed, Kothapally, 1999–2000 (Sreedevi et al. 2004)

Landform	Total input (kg ha^{-1})			Total output (kg ha^{-1})			Balance (kg ha^{-1})		
	N	P	K	N	P	K	N	P	K
Maize/pigeonpea BBF									
BBF	28	16	17	85	11	58	−57	5	−41
Flat	32	14	21	80	9	50	−48	5	−27
Sole maize									
BBF	21	10	0	75	14	71	−54	−4	−36
Flat	9	10	0	33	7	40	−24	3	−36

The technology was found efficient in Kothapally, and the results clearly revealed that the BBF system proved effective in conserving the rainwater, increasing the soil water in the profile and the winter sorghum grain yield by 18–26% and chickpea yield by 14–22% as compared to the flat sowing. The N, P and potassium (K) nutrient uptake by maize/pigeonpea intercropping system and sole maize was greater in the improved BBF system compared to that on the flat landform, which resulted in more crop yield on the BBF landform. The balances also showed that all systems were depleting N and K from soils and that more P is applied than removed by crops (Table 6.6) (ICRISAT 2002).

6.5.2 Contour Cultivation

Cultivation across the slope is simple method of cultivation, which can effectively increase rainfall infiltration and reduce runoff and soil loss on gently sloping lands. The contour cultivation involves performing cultural practices such as ploughing,

Table 6.7 Total productivity of sorghum and maize with boron and sulphur amendments at Adarsha watershed, Kothapally, 2001 (Sreedevi et al. 2004)

Treatment	Sorghum yield (kg ha^{-1})			Maize yield (kg ha^{-1})		
	Grain	Stalk	Total	Grain	Stalk	Total
Control	1460	2800	4260	1960	2360	4320
Boron (B)	1650	3030	4680	2360	2640	5000
Sulphur (S)	1890	3320	5210	2730	2840	5560
B+ S	1800	3490	5290	2580	3060	5640

planting and cultivating on the contours. It is observed that this system at the farmers' fields on Alfisols of Kabbalanala watershed near Bengaluru, India, increased soil moisture during cropping season from 35th to 43rd weeks over farmers' practice of up and down cultivation and reduced the runoff and soil loss and increased the yields of sesame, finger millet and groundnut. The furrows made either during planting time or during intercultural operations using traditional plough benefitted through increased as well as stabilized yield levels over years by 8–10%, apart from better rainwater management.

6.5.3   Integrated Nutrient Management

The integrated nutrient management (INM) approach was adopted to enable good crop growth with conserved soil and water. Detailed characterization of the soils showed that they were deficient in available phosphorus (P), N, zinc (Zn), boron (B) and sulphur (S). Amendments with B-, S- and B+S-treated plots resulted in 13–29% increase in sorghum grain yield and 20–39% increase in maize grain yield (Table 6.7) (ICRISAT 2002).

6.6   Integrated Pest Management

Integrated pest management (IPM) was adopted to optimize crop productivity along with integrated soil, water, crop and nutrient management in the watershed. The following IPM activities were implemented by the project, viz. crop surveys carried out revealed that farmers use chemical pesticides to control insect pests. *Helicoverpa* is the key pest on many crops. IPM practices such as use of pheromone traps, shaking of pigeonpea plants for controlling pod borers, use of pest-tolerant varieties and use of *Helicoverpa* nuclear polyhedrosis virus (HNPV) and bird perches were adopted (Sreedevi et al. 2004).

6.6.1 Village-Level HNPV Production

The project consortium identified farmer participants and initiated training in production, storage and usage of HNPV on different crops for minimizing pest damage. The farmers quickly adopted the technology and produced 2000 larval equivalent (LE) of HNPV and used it on cotton, pigeonpea and chickpea crops. ICRISAT supplied an additional 11,650 LE HNPV for use on these crops (Sreedevi et al. 2004).

6.7 Crop Simulation Models

The agricultural systems are very complex systems that are influenced by many interactions between biotic and abiotic factors. Abiotic factors like weather, soil properties, crop variety, planting date and spacing and inputs including irrigation and fertilizers and biotic stress like pests and diseases, weeds and soil-borne diseases determine crop status. A model is a mathematical representation of a real-world system. These models are very common in many disciplines, including the airplane industry, automobile industry, engineering, etc. However, in agricultural sciences these models are not extensively used. Development and use of simulation of crops started in the 1970s itself. A crop simulation model (CSM) is a mathematical model that describes processes of crop growth and development as a function of weather conditions, soil characteristics and crop management. Model simulates or imitates the behaviour of a real crop by predicting the growth of its components, such as leaves, roots, stems and grains. Thus, a crop growth simulation model not only predicts the final state of crop production or harvestable yield but also contains quantitative information about major processes involved in the growth and development of the crop. Modelling is able to explain the correlation among the components of complex systems, give more insight into processes and verify the consequences of management as well as explore the potential for modification. Besides, it is able to integrate a lot of information from various experiments at various sites and manage to extrapolate the information to another region of interest under various soil and climatic conditions.

Crop simulation models hold a vital place in the development of innovative crop management strategies and agricultural sustainability under a continuously changing climate, as it expresses the response of crops to meteorological, edaphic and biological factors. These models aid in decision-making, forecasting of crop growth and development, minimizing the yield gaps and selecting suitable genotypes and appropriate sowing dates for sustainable crop production under changing climatic scenarios. It is becoming a valuable tool for increasing our understanding of crop physiology and ecology for sustainable agricultural production. Models are used for yield prediction (Tsuji et al. 1994), simulation of crop damage, making policies, finding out interaction effects such as genotype by environment (G × E) interaction, soil moisture dynamics, nitrate losses, soil erosion and other factors related to agriculture. Crop models are typically applied for estimating potential yields and yield gaps between farmers and potential yields, yield forecasting, assessing impact of climate change and variability, optimization of crop management and genotypes and environmental interactions.

Many models have been developed, including Decision Support System for Agrotechnology Transfer (DSSAT), Agricultural Production Systems sIMulator (APSIM), Soil-Plant-Atmosphere System Simulation (SPASS), Root Zone Water Quality Model (RZWQM), Simulateur mulTIdisciplinaire pour les Cultures Standards (STICS), Soil-Water-Atmosphere-Plant environment (SWAP), World Food Study (WOFOST), Cropping System Simulation (CropSyst), etc. Among these models DSSAT and APSIM have been extensively applied globally to evaluate different cropping options and to assess impact of climate change and variability.

6.7.1 DSSAT Crop Models

The Decision Support System for Agrotechnology Transfer (DSSAT) was originally developed by an international network of scientists, cooperating in the International Benchmark Sites Network for Agrotechnology Transfer project (IBSNAT 1993; Tsuji 1998; Uehara 1998; Jones et al. 1998), to facilitate the application of crop models in a systems approach to agronomic research. The DSSAT helps decision-makers by reducing the time and human resources required for analysing complex alternative decisions. The DSSAT is a collection of independent software programs that operate together; crop simulation models are at its centre (Fig. 6.5). Databases describe weather, soil, experiment conditions and measurements and genotype information for applying the models to different situations. Software helps users prepare these databases and compare simulated results with observations to give them confidence in the models or to determine if modifications are needed to improve accuracy (Uehara 1989; Jones et al. 1998). The cropping system model (CSM) in DSSAT is structured in a modular format in which components separate long scientific discipline lines and have interfaces which allow replacement or addition of modules. CSM now incorporates all crop models as modules using a single soil module and a single weather module. The new cropping system model now contains models of 40+ crops derived from the original SOYGRO, PNUTGRO, CERES-Maize and CERES-Wheat crop growth models. Apart from crop modules, there are analysis models available in DSSAT.

6.7.1.1 Seasonal and Risk Analysis

Risk analysis is one of the main applications of DSSAT and allows users to evaluate alternate management practices for single growing seasons that account for both weather and economic uncertainty. Using the seasonal analysis option of DSSAT, a user can compare the interaction of genotype and management for different

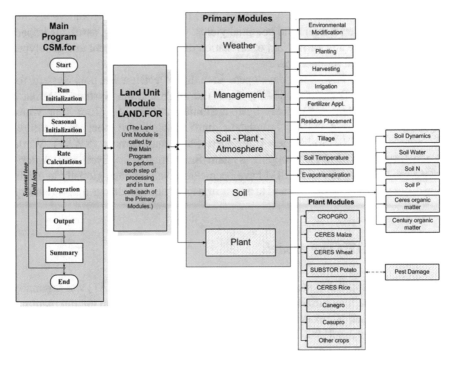

Fig. 6.5 Schematic diagram of DSSAT cropping systems models architecture (Source: www. DSSAT.Net)

environments, especially long-term historical weather data. Usually, a user defines at least two or more management scenarios to compare. Normally for the weather inputs, at least 30 years of historical weather data are selected, analogous to climate normals that are based on 30 years of historical weather data. The simulations are conducted for each unique combination of crop management and weather year. This provides a simulated distribution for yield, yield components and other simulated variables. The economic uncertainty can be defined through prices files.

6.7.1.2 Rotation and Long-Term Simulations

Cropping systems are not really defined by single growing seasons, but the long-term management practices that are implemented. This requires long-term simulations, starting with initialization of the cropping system environment with respect to soil water; nitrogen, phosphorus and potassium for the individual soil layers or horizon; soil surface residue; and soil organic matter. The sequence analysis module in DSSAT can be used to simulate crop rotations with historical weather data and performance of systems.

6.7.2 APSIM Models

The Agricultural Production Systems sIMulator (APSIM) was developed to simulate biophysical processes in agricultural systems, particularly as it relates to the economic and ecological outcomes of management practices in the face of climate risk. APSIM is developed by Agricultural Production Systems Research Unit (APSRU), CSIRO, Australia. APSIM is also being used to explore options and solutions for the food security, climate change adaptation and mitigation and carbon trading problem domains (Holzworth et al. 2014). From its inception 20 years ago, APSIM has evolved into a framework containing many of the key models required to explore changes in agricultural landscapes with capability ranging from simulation of gene expression through to multi-field farms and beyond.

APSIM software has been developed on modular approach. Modules can be biological, environmental, managerial or economic and are linked via the APSIM "engine". The "engine" is a communication system that passes information between modules according to a standard protocol (Fig. 6.6). The fact that two modules are not directly linked allows modules to be plugged in or pulled out of the "engine" depending on the specifications for the simulation task. In this way, the simulation capacity of APSIM is limited only by the availability of modules to simulate the processes peculiar to the system of interest. APSIM is structured around plant, soil and management modules. These modules include a diverse range of crops, pastures and trees, soil processes including water balance, N and P transformations, soil pH, erosion and a full range of management controls. APSIM resulted from a need for

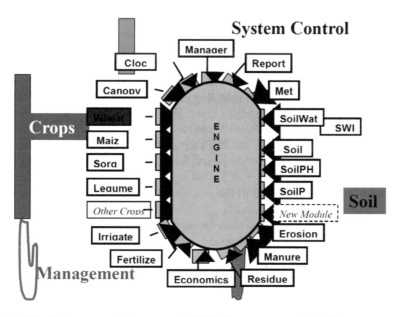

Fig. 6.6 Plug-in/plug-out modular system of APSIM (Source: www.apsim.info)

tools that provided accurate predictions of crop production in relation to climate, genotype, soil and management factor while addressing the long-term resource management issues.

APSIM has been used in a broad range of applications, including support for on-farm decision-making, farming systems design for production or resource management objectives, assessment of the value of seasonal climate forecasting, analysis of supply chain issues in agribusiness activities, development of waste management guidelines, risk assessment for government policy making and as a guide to research and education activity. Various modules of APSIM have been evaluated globally with observed datasets. Maize and cowpea intercropped under a range of soil water and fertility conditions, and with the cowpea planted at different times relative to the maize planting time, have been evaluated and found to be satisfactory (Carberry et al. 2002). Ludwig and Asseng (2006) found that in the southern, higher rainfall part of south-western Australia, yield and gross margin will increase for all likely future climate scenarios. In the drier part of the region, negative effects of 15% reduced rainfall can be compensated for by a 2 °C increase in temperature and 50% higher CO_2 concentrations. In south Australia, median grain yield is projected to decrease across all locations from 13.5% to 32% under the most likely climate change scenarios (Luo et al. 2005).

6.8 General Circulation Models (GCMs)

Coupled atmosphere-ocean general circulation models (GCMs) simulate different realizations of possible future climates at global scale under contrasting scenarios of land-use and greenhouse gas emissions. Outputs from many GCMs are available in the public domain, notably in the World Climate Research Programme's (WCRP's) Coupled Model Intercomparison Project Phase 5 (CMIP5) multi-model dataset. This dataset contains model outputs from 22 of the GCMs used for the fifth assessment and for a range of scenarios including the four scenarios reported in the IPCC's working group 1 of fifth assessment report (IPCC 2014). A set of four pathways were produced that lead to radiative forcing levels of 8.5, 6.0, 4.5 and 2.6 Wm^{-2}, by the end of the century. Each of the RCPs covers the 1850–2100 period, and extensions have been formulated for the period thereafter (up to 2300). The four RCPs formulated are RCP8.5, throughout the twenty-first century before reaching a level of about 8.5 W m^{-2} at the end of the century. In addition to this "high" scenario, there are two intermediate scenarios, RCP4.5 and RCP6.0, and a low so-called peak-and-decay scenario, RCP2.6, in which radiative forcing reaches a maximum near the middle of the twenty-first century before decreasing to an eventual nominal level of 2.6 W $m^{-2.}$ Many of the GCMs also project that the floods and droughts are going to be more frequent in the future.

The output data from GCM requires several additional processing steps before it can be used to drive impact models. Spatial downscaling, typically by regional climate models (RCM), and bias correction are two such steps that have already been addressed for. Yet, the errors in resulting daily meteorological variables may be too large for specific model applications. Crop simulation models are particularly sensitive to these inconsistencies and thus require further processing of GCM-RCM outputs. GCM models simulate typically a single time series for a given emission scenario. To help in agricultural policy making, data on near- and medium-term decadal time scale is mostly used, e.g., 2030 or 2050.

The earth system model CESM1 (version 1.2.1) is a fully coupled global climate model developed by National Center for Atmospheric Research (NCAR) in Boulder, Colorado. Many physics-based models serve for the different Earth system components. The atmosphere component, CAM5, provides a set of physics parametrizations and several dynamical cores, which also include advection (Baumgaertner et al. 2016). Base data has been taken from WorldClim version 1.4 for the period 1960–1990 (WorldClim 2005). Downscaled monthly climate projections at 10′ resolution for the climate model CESM1-CAM5 and the emission scenarios RCP 8.5 and RCP 4.5 for the period 2030 and 2050 were downloaded from CCAFS Data Centre (CCAFS 2018). Monthly climate projections for a representative pixel for Kothapally station have been extracted from the climate surface using ARCGIS 10.4 software. The temperature increases and rainfall changes have been derived by subtracting base period data from projected. Climate change projections for Kothapally have been provided in Table 6.2. Temperatures increase in range of 0.9–1.5 °C and 0.1–2.2 °C as per RCP4.5 and RCP 8.5, respectively, during 2030. Similarly, during 2050 temperatures increase in the range of 1.0–2.1 °C and 1.4–3.0 °C as per RCP4.5 and RCP 8.5, respectively. In both the scenarios annual rainfall has been increased in the range of 37–120 mm. In general rainfall increase during post-rainy season is higher Tables 6.8.

Table 6.8 Projected climate changes at Kothapally, Telangana, as per CESM1-CAM5 model

Parameter	Jan	Feb	Mar	Apr	May	Jun	Jul	Aug	Sep	Oct	Nov	Dec
Climate change scenario: RCP4.5, period, 2030												
Rainfall (mm)	1.0	1.0	6.0	4.0	−3.0	−4.0	−1.0	9.0	21.0	1.0	5.0	7.0
Tmax (°C)	0.9	1.3	0.8	1.0	1.4	1.1	1.3	0.8	0.6	1.1	1.5	0.8
Tmin (°C)	1.3	1.2	1.1	0.9	0.7	0.7	1.0	0.6	0.7	1.1	1.4	1.5
Climate change scenario: RCP4.5, period, 2050												
Rainfall (mm)	4.0	0.0	6.0	1.0	8.0	24.0	−28.0	50.0	26.0	31.0	10.0	7.0
Tmax (°C)	1.3	1.7	1.1	1.9	1.7	2.0	1.6	1.2	1.0	1.4	1.7	1.3
Tmin (°C)	1.9	1.4	1.4	1.1	1.1	1.5	1.3	1.0	1.2	1.7	2.1	1.9
Climate change scenario: RCP8.5, period, 2030												
Rainfall (mm)	4	4	4	0	16	4	−13	−13	5	27	6	21
Tmax (°C)	0.1	0.4	0.7	1.2	0.8	1.2	1.1	1.0	0.7	1.0	1.5	0.8
Tmin (°C)	1.4	0.8	0.9	0.7	0.6	0.9	1.0	0.9	0.9	1.3	2.1	2.2
Climate change scenario: RCP8.5, period, 2050												
Rainfall (mm)	1	2	4	−4	−2	−15	7	1	42	47	4	11
Tmax (°C)	1.4	2.3	2.0	2.6	3.0	3.5	2.2	1.8	1.5	1.8	2.2	1.4
Tmin (°C)	2.3	2.6	2.4	2.1	2.3	2.5	2.1	1.6	1.7	2.4	3.0	3.0

6.9   Data and Methods

Four major crops cultivated in the region, viz. cotton, maize, sorghum and pigeon-pea, were selected to study the impact projected climate change using crop simulation models. To run crop simulation models, minimum datasets for the site are required. The details of the minimum dataset were described by Jones et al. (2003). Site characters (latitude, longitude and elevation) have been collected by using a GPS. The weather parameters required by the model include precipitation minimum and maximum temperatures and solar radiation. An automatic weather station has been established in Kothapally, during 1999 as part of their watershed activities. Daily and hourly weather data is being recorded in this weather station. Weather data from 1998 to 2007 (20 years) has been used in this study. Weather data for the year 1998 has been taken from weather station at ICRISAT, Patancheru, which is 30 km away, and data for 1999–2017 has been used from weather station at Kothapally. Projected climate data is derived by adding monthly temperature increase to the daily observed data of respective months. For rainfall a factor has been derived by dividing projected rainfall by base rainfall for each month. To derive future projected rainfall for Kothapally, daily rainfall in particular month is multiplied by the factor of respective month.

The soil profile characteristics required by layer are saturation limit, drained upper and lower limit of water availability, bulk density, organic carbon, pH, root growth factor, runoff and drainage coefficients. Soil samples were collected to estimate soil physical properties. Soil parameters needed for simulation were estimated from the soil survey data using the S Build program available in DSSAT v4.7. Soil profile characteristics used in the model are provided in Table 6.3. Required crop management data include cultivar, sowing date, plant population, row spacing, sowing depth and dates and amounts of irrigation, fertilizers applied and other inter-cultivation activities. All standard crop management practices established for the region were followed in the simulation. Sagar et al. have developed genetic coefficients for the cultivar RCH 791 based on the field experiments and found the results were well in the range observed yield. In the present study the cotton cultivar RCH-791 has been used. For sorghum simulation cultivar CSV 15 has been used. The cultivar coefficients for sorghum have been developed using experiments conducted at ICRISAT (Singh et al. 2014). Maize is simulated using cultivar HT 5402. The cultivar coefficients information is given in Table 6.4. This cultivar was validated by the field experiments. Impact of projected climate change on pigeonpea is estimated using APSIM model. ICRISAT has conducted field experiments to generate genetic coefficients of pigeonpea cultivar ICPL87119 under APSIM during 2013–2015. The cultivar coefficients developed by ICRISAT for ICPL87119 are used in this study (Tables 6.9, 6.10).

Table 6.9 Soil properties used for simulation

Soil depth	Lower limit	Upper limit	SAT SW	EXTR SW	INIT SW	Root dist	Bulk density	pH	OC
Cm	cm3/cm3	cm3/cm3	cm3/cm3	cm3/cm3	cm3/cm3		g/cm3		%
0–5	0.292	0.407	0.422	0.115	0.292	0.5	1.39	7.9	0.59
5–15	0.292	0.407	0.422	0.115	0.292	0.5	1.39	7.9	0.59
15–30	0.300	0.413	0.428	0.113	0.300	0.2	1.39	7.9	0.50
30–45	0.300	0.413	0.428	0.113	0.300	0.2	1.39	7.9	0.50
45–60	0.294	0.405	0.420	0.111	0.294	0.1	1.38	8	0.41
60–75	0.294	0.405	0.420	0.111	0.294	0.1	1.38	8	0.41
75–90	0.193	0.289	0.355	0.096	0.193	0.1	1.54	8	0.30

Table 6.10 Cultivar coefficients used in DSSAT simulations for Kothapally

Cotton: RCH 791		Maize: HT 5402		Sorghum: CSV 15	
Parameter	Value	Parameter	Value	Parameter	Value
CSDL	23	P1	285	P1	400
PPSEN	0.01	P2	0.5	P2	102
EM-FL	46	P5	780	P2O	12.8
FL-SH	14	G2	900	P2R	120
FL-SD	19	G3	10.5	PANTH	617.5
SD-PM	56	PHINT	75	P3	152.5
FL-LF	75			P4	81.5
LFMAX	1.3			P5	640
SLAVR	420			PHINT	49
SIZLF	410			G1	7
XFRT	0.91			G2	6.2
WTPSD	0.18				
SFDUR	40				
SDPDV	30				
PODUR	12				
THRSH	91				
SDPRO	0.145				
SDLIP	0.13				

6.9.1   Simulation of Selected Crop Yields Under Climate Change Scenarios

To simulate yield of selected crops under selected climate change scenarios, various models under DSSAT and APSIM were used. For estimating impact of climate change on sorghum and maize, the CERES models available in DSSAT V4.7 were used. For cotton simulation CROPGRO model available in DSSAT V4.7 has been used. For running multiyear runs, seasonal analysis programme available in DSSAT v4.7 was used. Simulations were carried out for the years 1998–2017 (20 years). The simulations were initiated on 15 May each year and the soil profile was considered to be at the lower limit of water availability (SLL) on that day. The sowings were done on 25 June every year considering the fact that the sufficient rains are available for sowing by that time. A plant population of 18 plants m^{-2} for sorghum with the rows pacing of 30 cm and ten plants m^{-2} with row spacing of 45 cm have been used. For cotton simulation a plant population of 15 plants m^{-2} with row spacing of 90 cm has been used. The following treatments (scenarios) for each crop comprising of present and future climate change projections were evaluated.

1. Present (present weather data 1998–2017)
2. CESM1-CAM5-RCP85–2030 (projected weather with GCM CESM1-CAM5 for RCP 8.5 scenario and for the year 2030)
3. CESM1-CAM5-RCP85–2050 (projected weather with GCM CESM1-CAM5 for RCP 8.5 scenario and for the year 2050)
4. CESM1-CAM5-RCP45–2030 (projected weather with GCM CESM1-CAM5 for RCP 4.5 scenario and for the year 2030)
5. CESM1-CAM5-RCP45–2050 (projected weather with GCM CESM1-CAM5 for RCP 4.5 scenario and for the year 2050)

6.10 Impact of Projected Climate Change on Crop Yields

Climate change in future will alter the growing conditions of crops due to increase in temperature and change in the rainfall patterns. In the semi-arid tropics, the duration of growing seasons (water availability period) will generally decrease, and the frequency of abiotic (particularly temperature and drought stresses) and biotic stresses is most likely to increase. Such adverse growing conditions in future will impact the crop yields negatively. Overall impact of climate change on crop yields is determined by the current climate and soil characteristics of the site and the future projected climate change. The vulnerability of the agriculture sector to both climate change and variability is well established. The general consensus is that changes in temperature and precipitation will influence plant growth and crop yield (Ma et al. 2013). In many developing countries, climate change is also expected to lead to changes in farming systems and will put more pressure on the rural community to cope with these changes and to build up their adaptive capacities (Liwenga 2008; Deng 2010). Mall et al. (2004) predict that yield of soybean in India could decrease in the range of 10–20% under various climate change scenarios. A study conducted by Mishra et al. (2013) on the rice and wheat using DSSAT based on the projections by a regional climate model in Indo-Gangetic plains found that there were reduction yield of wheat and rice in upper Gangetic plains while increase in lower Gangetic plains. In the present study, in all the four climate change scenarios, rainfall during June and July is decreasing and increasing during August to November. Due to increasing rainfall during August to November, long-duration crops like cotton and pigeonpea have benefitted and sustained their yields. Short-duration crops like maize and sorghum have been impacted negatively because of reduced rainfall during June and July.

6.10.1 Climate Change Impacts on Cotton

In this study, the mean yield and total biomass of cotton were 3507 kg ha^{-1} and 5623 kg ha^{-1}, respectively, under baseline climate (Tables 6.6 and 6.7). Under climate change scenario RCP4.5 by the year 2030, biomass increased by 1% and grain yield by 4%. In general, projected increase in temperatures shortens various phenological stages and total crop duration of crops which impact negatively on the performance of the crop. In the case of cotton crop, days to anthesis (flowering) reduced by 2 days and days to maturity reduced by 10 days (Tables 6.8 and 6.9). Negative impact of increased temperature was compensated by increased rainfall. As per the model CASM1-CAM5 under the scenario RCP4.5 by the year 2050, grain yield increased by 10% and total biomass increased by 6%. Durations for flowering and maturity are shortened by 3 and 14 days, respectively. Climate change projections by the GCM model CESM1-CAM5 under RCP 8.5 by the year 2030, biomass and grain yield of cotton increased by 4% and 6%, respectively. Durations for flowering and maturity were reduced by 2 and 11 days, respectively (Tables 6.14 and 6.15). Similarly, by applying GCM model CESM1-CAM5 under climate change scenario RCP 8.5 by the year 2050, cotton crop yield is expected to increase by 7% and total biomass by 4%. Higher projected increased temperature causes to shorten the crop duration by 18 days compared to base. Year to year variability cotton yield is also reduced due to increased rainfall in the post-rainy season (Fig. 6.7). Increased rainfall and increased temperatures resulted into more usage of water by the crop (Table 6.11). Increased rainfall also caused more runoff which indicates potential for the water harvesting.

6.10.2 Climate Change Impacts on Maize

Maize is the third most important food crops of India after rice and wheat and contributes nearly 9% in the national food basket. Maize is cultivated in 9.6 million ha with productivity of 2.6 t ha^{-1} in India. The predominant maize-growing states are Karnataka, Madhya Pradesh, Maharashtra, Rajasthan, Telangana, Uttar Pradesh, Jharkhand and Andhra Pradesh. Most of the maize-growing areas fall in semi-arid climate zone and mostly cultivated under rain-fed conditions. The area under maize in India is slowly growing in recent years. Since maize is cultivated under rain-fed conditions, they are more vulnerable to climatic variation. Demand for maize would be double compared to present in the developing countries by the year 2050 (Rosegrant

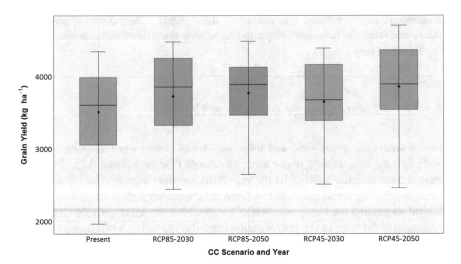

Fig. 6.7 Impact of climate change on yield of cotton at Kothapally

Table 6.11 Cultivar coefficients used for pigeonpea simulation in APSIM for Kothapally

Parameter	Units	Value/range
x_pp_hi_incr	h	24
y_hi_incr	1/day	0.004
x_hi_max_pot_stress		0.0 1.0
y_hi_max_pot		0.15 0.15
cumvd_emergence	d	0100
tt_emergence	oCd	272.0272.0
est_days_emerg_to_init	d	40
x_pp_end_of_juvenile	h	11.4 13.2 13.3
y_tt_end_of_juvenile		12,250 100,000
x_pp_floral_initiation	h	1 24
y_tt_floral_initiation	oCd	123.0123.0
x_pp_flowering	h	1 24
y_tt_flowering	oCd	90.0 90.0
x_pp_start_grain_fill	h	1 24
y_tt_start_grain_fill	oCd	636.0636.0
tt_end_grain_fill	oCd	48
tt_maturity	oCd	48
x_stem_wt	g	0 4 9 25 85,130
y_height	Mm	06001000 1300 2000 2100

et al. 2008).The spatio-temporal variations in projected changes in temperature and rainfall are likely to lead to differential impacts in the different regions. In particular, monsoon yield is reduced most in Southern Plateau up to 35% (Byjesh et al. 2010).

As per the GCM CESM1-CAM5 model, at Kothapally, the grain yield of maize in both the climate change scenarios, by 2030 and 2050, is decreasing. Under baseline climate potential biomass and yield estimated were 10,456 and 6072 kg ha^{-1}, respectively. Since maize crop duration is about 110 to 120 days, the impact of reduced rainfall in June and July and increased temperatures is clearly visible. As per climate change scenario RCP4.5 by the year 2030, biomass decreased by 1% and grain yield decreased by 5%. Duration of anthesis is shortened by 2 days and days to maturity reduced by 5 days (Tables 6.8 and 6.9). As per the model CASM1-CAM5 under the scenario RCP4.5 by the year 2050, grain yield decreased by 9% and total biomass decreased by 4%. Durations for flowering and maturity are shortened by 5 and 9 days, respectively. As per climate change projections by the GCM model CESM1-CAM5 under RCP 8.5 by the year 2030, biomass and grain yield of cotton decreased by 7% and 2%, respectively. Durations for flowering and maturity were reduced by 4 and 7 days, respectively (Tables 6.9 and 6.10). Similarly by applying GCM model CESM1-CAM5 under climate change scenario RCP 8.5 by the year 2050, maize crop yield is expected to decrease by 12% and total biomass by 4%. Higher projected increased temperature causes to shorten the crop duration by 10 days compared to base. Year to year variability of grain yield of maize is also increased in all the climate change scenarios (Fig. 6.8). Increased temperature causes to shorten the crop duration and crop water uptake. Increased rainfall during August compensated the deficit water requirement at the end of the crop (Table 6.11).

6.10.3   Climate Change Impacts on Sorghum

Sorghum is one of the major cereal crops grown mostly under rain-fed condition, and it continued to be main staple food for marginal farmers of developing countries in Asia and Africa (Murthy et al. 2007). Due to its higher drought tolerance over other cereal crops, sorghum is highly suitable to semi-arid tropic (SAT) crop production system (Ludlow and Muchow, 1990). In India sorghum is cultivated in 2.1 million ha with production of 1.96 million t during kharif and 3.6 million ha with production of 2.6 million t during rabi season (DACNET 2016). The sorghum productivity in India is far low (864 kg ha^{-1}) compared to global average (1481 kg ha^{-1}) (FAOSTAT 2018). Grain sorghum yield is mostly influenced by crop management practices, variation in rainfall amount and its distribution, soil water content at planting and soil water-holding characteristics (Assefa et al. 2010). Sandeep et al. (2017) have reported increased water requirement for sorghum over majority of sorghum-growing regions both in *kharif* and *rabi* season during 2050 to 2080. 2016. Grain yield of sorghum decreased under rain-fed and no-stress conditions using various GCM output data across the Indian locations (Gangadhar Rao et al. 1995).

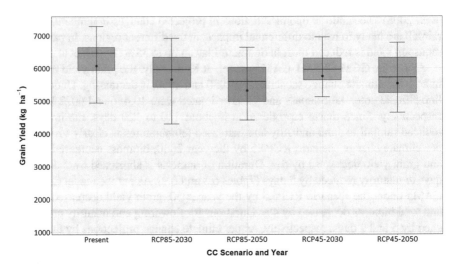

Fig. 6.8 Impact of climate change on yield of maize at Kothapally

Sorghum grain yield at Kothapally has decreased as per the GCM CESM1-CAM5 model, in both the climate change scenarios RCP4.5 and RCP8.5, by 2030 and 2050. Under baseline climate, potential biomass and yield of sorghum estimated were 8332 and 3585 kg ha^{-1}, respectively. Decreased projected rainfall during crop-growing season of sorghum has impacted negatively. As per climate change scenario RCP4.5 by the year 2030, biomass increased by 6%, whereas there is no change in grain yield. Duration of anthesis is shortened by 3 days and days to maturity reduced by 5 days (Tables 6.9 and 6.10). As per the model CASM1-CAM5 under the scenario RCP4.5 by the year 2050, grain yield decreased by 9% and total biomass decreased by 5%. Durations for flowering and maturity are shortened by 5 and 7 days, respectively. As per climate change projections by the GCM model CESM1-CAM5 under RCP 8.5 by the year 2030, biomass and grain yield of sorghum decreased by 6% and 3%, respectively. Durations for flowering and maturity were reduced by 5 days each (Tables 6.14 and 6.15). Similarly, by applying GCM model CESM1-CAM5 under climate change scenario RCP 8.5 by the year 2050, sorghum crop yield decreased by 13% and total biomass by 10%. Higher projected increased temperature causes to shorten the crop duration by 10 days compared to base. Year to year variability of grain yield of sorghum is also increased in all the climate change scenarios (Fig. 6.9). Increased temperature shortens the crop duration and decreases crop water uptake (Table 6.16). Increased rainfall during August compensated the deficit water requirement at the end of the crop (Table 6.16).

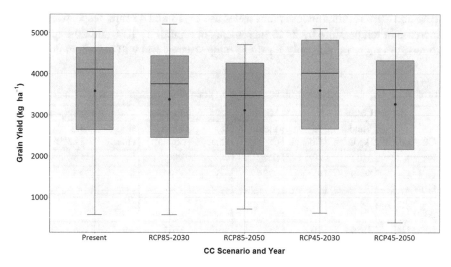

Fig. 6.9   Impact of climate change on yield of sorghum at Kothapally

6.10.4   Climate Change Impacts on PigeonPea

India is the largest producer (17–18 m t), consumer (22–23 m t) and importer (4–5 m t) of pulses till 2016–17. Chickpea, lentil and pigeonpea account for 39%, 10% and 21%, respectively, of the total pulse production in India (NFSM 2009). Pigeonpea and chickpea are major pulse crops, which contribute about 60% of total pulse production in India. Climate changes have a major impact on rain-fed crops including pulses (Basu et al. 2009). Pigeonpea (*Cajanus cajan*) is a major source of protein supplement for most Indian population. Presently there is a wide gap between farmers' yield and potential yield in all the agroecological regions of India (Bhatia et al. 2006). Birthal et al. (2014) estimated that yield of pigeonpea could be reduced in the rage of 7–25% due to climate change by the end of century.

In a study conducted at ICRISAT to assess the climate change impacts on pigeonpea with the variety TS-3R at Kalaburagi, Karnataka, ten climate change scenarios and the present climate conditions were considered using the calibrated APSIM model (Kesava Rao et al. 2013). The scenarios included 1 °C and 2 °C increase in both maximum and minimum temperatures with 10% and 20% decrease and increase in rainfall. Results showed that in Kalaburagi, increase in temperature by 2 °C could reduce pigeonpea yields by about 16%. Rainfall decrease of 10% from present coupled with 2 °C increase in temperature could reduce yields further by 4%, making the total reduction to be at 20%. Increased temperature could shorten the crop duration. Days to flowering shortened by 2 and 4 and the total crop duration by 5 and 9 days with increase in temperature by 1 and 2 °C, respectively. Increase in temperature causes more transpiration per day which results in water stress during the dry periods. Water balance outputs have shown that decrease in rainfall by

10% and 20% resulted in less plant water use by 18 and 45 mm, respectively, with increase in temperature by 2 °C. Increments in rainfall by 10% and 20% are likely to result in more rainfall only for those rainy seasons and will not affect non-rainy

Table 6.12 Impact of projected climate change on grain yield of major crops at Kothapally

	Cotton		Maize		Sorghum		Pigeonpea	
CC scenario	Yield (kg ha^{-1})	% Change	Yield (kg ha^{-1})	% Change	Yield (kg ha^{-1})	% change	Yield (kg ha^{-1})	% change
Present	3507	–	6072	–	3585	–	1189	–
CESM1-CAM5-RCP45–2030	3648	4	5754	−5	3596	0	1225	3
CESM1-CAM5-RCP45–2050	3856	10	5531	−9	3264	−9	1266	6
CESM1-CAM5-RCP85–2030	3725	6	5650	−7	3380	−6	1258	6
CESM1-CAM5-RCP85–2050	3769	7	5320	−12	3119	−13	1246	5

Table 6.13 Impact of projected climate change on total biomass of major crops at Kothapally

	Cotton		Maize		Sorghum		Pigeonpea	
CC scenario	Biomass (kg ha^{-1})	% change	Biomass (kg ha^{-1})	% change	Biomass (kg ha^{-1})	% change	Biomass (kg ha^{-1})	% change
Present	5623	–	10,456	–	8332	–	7927	–
CESM1-CAM5-RCP45–2030	5672	1	10,336	-1	8801	6	8165	3
CESM1-CAM5-RCP45–2050	5983	6	10,076	−4	7909	−5	8440	6
CESM1-CAM5-RCP85–2030	5825	4	10,224	−2	8106	−3	8391	6
CESM1-CAM5-RCP85–2050	5832	4	10,012	−4	7537	−10	8308	5

days. Thus, additional rainfall has contributed more towards runoff and drainage than evapotranspiration. Shantanu Kumar et al. (2011) found that in Bundelkhand region, 0.1 °C caused the yield reduction of pigeonpea up to 22 kg ha^{-1} and 10 mm of rainfall caused the reduction of 8 kg ha^{-1}.

Mean grain yield and total biomass of pigeonpea were 1189 kg ha^{-1} and 7927 kg ha^{-1}, respectively, under baseline climate (Tables 6.12 and 6.13). Under climate change scenario RCP4.5 by the year 2030, biomass and grain yield increased by 3% each. Under baseline climate days to anthesis is 119 days and maturity is 184 days. The duration of anthesis and maturity reduced by 2 days and 9 days, respectively, under the scenario of RCP4.5 by the year 2030. (Tables 6.8 and 6.9). Negative impact of increased temperature was compensated by increased rainfall. As per the model CASM1-CAM5 under the scenario RCP4.5 by the year 2050, grain yield and total biomass by 6% each. Durations for flowering and maturity are shortened by 9 and 22 days, respectively. Under climate change projections by the GCM model CESM1-CAM5 under RCP 8.5 by the year 2030, biomass and grain yield increased by 6% each. Duration for flowering and maturity were reduced by 2 and 11 days, respectively (Tables 6.14 and 6.15). Similarly, by applying GCM model CESM1-CAM5 under climate change scenario RCP 8.5 by the year 2050, pigeonpea crop yield and biomass are expected to increase by 5%. Higher projected increased temperature causes to shorten the crop duration by 17 days compared to base. Year to year variability of grain yield of cotton is also reduced due to increased rainfall in the post-rainy season (Fig. 6.10). Increased rainfall and temperatures

Table 6.14 Impact of projected climate change on duration for anthesis of major crops at Kothapally

CC scenario	Cotton	Maize	Sorghum	Pigeonpea
Present	61	72	77	119
CESM1-CAM5-RCP45–2030	59	69	74	117
CESM1-CAM5-RCP45–2050	58	67	72	110
CESM1-CAM5-RCP85–2030	59	68	73	117
CESM1-CAM5-RCP85–2050	57	66	70	115

Table 6.15 Impact of projected climate change on duration for maturity of major crops at Kothapally

CC scenario	Cotton	Maize	Sorghum	Pigeonpea
Present	171	116	118	184
CESM1-CAM5-RCP45–2030	161	111	113	175
CESM1-CAM5-RCP45–2050	157	109	111	162
CESM1-CAM5-RCP85–2030	159	110	113	173
CESM1-CAM5-RCP85–2050	153	106	108	167

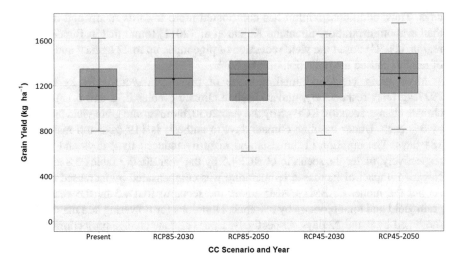

Fig. 6.10 Impact of climate change on yield of pigeonpea at Kothapally

Table 6.16 Impact of projected climate change on duration for evapotranspiration (mm) of major crops at Kothapally

CC scenario	Cotton	Maize	Sorghum	Pigeonpea
Present	480	384	369	435
CESM1-CAM5-RCP45–2030	492	384	374	459
CESM1-CAM5-RCP45–2050	505	384	369	458
CESM1-CAM5-RCP85–2030	495	382	369	460
CESM1-CAM5-RCP85–2050	493	369	354	470

Table 6.17 Impact of projected climate change on duration for runoff (mm) of major crops at Kothapally

CC scenario	Cotton	Maize	Sorghum	Pigeonpea
Present	132	136	137	79
CESM1-CAM5-RCP45–2030	139	143	144	80
CESM1-CAM5-RCP45–2050	182	183	184	103
CESM1-CAM5-RCP85–2030	140	142	143	78
CESM1-CAM5-RCP85–2050	161	147	160	99

resulted into more usage of water by the crop by 35 mm (Table 6.16). Runoff increased in the range of 2 and 24 mm due to increased rainfall under climate change scenarios which indicates potential for water harvesting (Tables 6.17).

6.11 Adaptation Strategies

The watershed development programme provided tangible economic benefits to individuals through an integrated approach. It focused on natural resource management, improved cultivars and IPM on soil and moisture conservation, water harvesting and afforestation. The benefits from contour bunding, check dams, percolation tanks, gabion structures and gully plugs are communal, and their impact is not immediately visible to the farmers. In integrated watershed development, there is large scope for natural resources management and multidisciplinary approach. The interest of the individual farmers and the community is the driving principle for design and development of technologies. In this approach, in situ water conservation, field bunds, soil management, land preparation and vegetative bunds are some of the interventions initiated. The benefits of conserving soil moisture, augmenting soil fertility through soil management, etc., show immediate visible gains to farmers in the form of higher yields and reduced input costs (Wani et al. 2002). Integrated watershed management requires a holistic enhancement of biophysical and human resources in the village rather than a mere soil and water conservation programme (Wani et al. 2002).

Natural resource management in isolation from livelihood and production activities does not bring optimized benefits, but when coupled with human resource development, improved varieties and water management bring in the desired gains on a more sustainable basis. Hence, a holistic integrated watershed management programme was initiated to reduce resource degradation and improve the livelihoods of the poor. In Adarsha watershed the starting point has been to recognize the needs of individual farmers. During an interview with the farmers in Kothapally, farmers attributed the higher impact of the programme to the tangible benefits realized by individual farmers. The basic lacunae in the participation of small farmers were addressed through the approach of emphasizing on-farm interventions that improved crop yields and incomes for the individual farmers. The sense of community ownership, individual achievement in tackling the long-standing problems of drought and resource degradation and the private economic benefits ensured enhanced individual participation.

In the preparation of the micro-plan, emphasis was given to individual benefits. The problem of general reluctance of the community to engage in watershed management when benefits are delayed and intangible was addressed by providing integrated soil and water management technologies that provide immediate benefit to farmers. This has stimulated their interest and built the foundation for collective action and sustainable community resource management. For example, the BBF system of land preparation conserved soil and retained soil moisture in situ thereby benefiting the farmers, while draining out excess water during heavy rains which was again harnessed in community-based check dams. Both the individual short-term and long-term community benefits were evenly balanced in the integrated watershed management programme.

With regard to post-project sustainability, the farmers have started developing strategies. The user groups (UGs) which were formed for each water storage structure and gained benefits through these structures will form interest groups and will take up desiltation and maintenance in the future. They may transform into common interest groups or SHGs in place of UGs. They also look towards using the watershed development fund (WDF) as the revolving fund so that the fund is not depleted at any time. The women SHGs who have started local enterprises in the production of vermicompost, preparation of HNPV, etc., will continue to expand and diversify these activities. The remarkable progress made in the implementation of a new science-based farmer participatory consortium modelled by ICRISAT is making Adarsha watershed a promising model in watershed management. The example of Adarsha watershed is reaching different states in India as well as other countries like Vietnam and Thailand in Asia. International donors are now asking to replicate and scale up this model in new areas. Already scaling up of this approach in selected watersheds of Kurnool, Nalgonda and Mahbubnagar districts of Andhra Pradesh is done with the support of the Department for International Development (DFID), UK, and through the Andhra Pradesh Rural Livelihoods Programme (APRLP). Similarly, the Sir Dorabji Tata Trust, Mumbai, and the Asian Development Bank (ADB) have provided funds to replicate this approach in selected watersheds in India (Madhya Pradesh, Rajasthan and Gujarat), Thailand and Vietnam.

6.12   Summary

Drought-prone areas are categorized by land degradation, low and erratic rainfall, low rainwater use efficiency, high soil erosion, inherently less fertile soils and subsistence agriculture. The farmers in these areas are very poor and their ability to take risk and invest in necessary inputs for optimizing production is low. There is a general tendency to exploit groundwater for food crops by the few resourceful farmers. Dryland areas are repeatedly prone to drought because of their geographical location. Also these areas are prone to waterlogging situations during the cropping season due to torrential downpours interspersed with long dry spells.

Watershed programmes implemented in India for improving the productivity in drought-prone areas have mainly focused on natural resource conservation and interventions such as soil and water conservation and to some extent afforestation in the government forestlands. Sufficient emphasis and efforts were not targeted to build the capacity of the community for enhanced management of the resource base while improving the livelihoods of the poor. Similar issues like gender equity and benefits for the landless have not been addressed adequately, thereby resulting in a mere water storage structure-driven investment giving only wage labour benefits to some deprived sections of the society. The watershed projects should move from purely soil and moisture conservation and water harvesting interventions to a wholesome community-based integrated watershed management approach which creates a voice and stake for the landless and poor women and men. Also, it is necessary to involve the primary stakeholders right from the beginning and build up their capacities to take the programme towards a sustainable initiative. The project design and proposed intervention should also aim at building local capacity for sustainable management of the resource base especially in the post-project phase.

A new science-based farmer participatory consortium model for efficient management of natural resources emerged from the lessons learned from long-term watershed management research by ICRISAT along with the national partners like CRIDA, NRSA and DWMA. This new approach was implemented in Adarsha watershed, Kothapally. The important components of the new model, which are distinctly different from earlier models, are:

- a consortium of institutions which provides technical back stopping and essential advisory services for community watershed development facilitated through experienced NGOs; greater role for farmers and local communities in project design, implementation, monitoring and evaluation; no subsidy (users pay principal for interventions on private lands); low-cost soil and water conservation structures; in situ conservation measures on farmers' fields to ensure tangible economic benefits to individuals; interventions that enhance the productivity of traditional crops and provide livelihood benefits to the poor and landless farmers; emphasis on capacity building of the stakeholders to become trainers; and continuous monitoring and refinement jointly by farmers and other partners.

For conservation of soil and water, the following community-based interventions were implemented using watershed development funds: water storage structures, gully control structures, mini percolation pits and gabion structures. Similarly, farmer-based soil and water conservation measures like BBF and contour planting to conserve in situ soil and water; use of tropicultor for planting, fertilizer application and weeding operation; and field bunding and planting of *Gliricidia* on field bunds for strengthening and conserving rainwater and supply of N-rich organic matter were implemented in individual farmers' fields.

To enable good crop growth from conserved soil and water, INM practices such as use of inorganic and organic nutrients, application of deficient micronutrients like S and B and balanced application of all the essential nutrients were advocated. For effective control of pests and diseases, the consortium initiated training on production, storage and usage of HNPV. The village common lands and wastelands were planted with custard apple saplings, *Gliricidia* saplings and avenue plantations as a part of the village afforestation programme. The women SHGs were motivated to take up vermicomposting as a microenterprise to provide biofertilizers on local demand and generate income. The implementation of soil and water conservation interventions resulted in about 30–45% reduction in runoff and rise in the groundwater level. Due to additional groundwater recharge, a total of about 200ha in post-*kharif* season and about 100ha in post-*rabi* season are cultivated with different crops and cropping sequences. Adoption of improved practices like high-yielding cultivars and integrated nutrient and pest management practices by farmers in the Adarsha watershed resulted in increased productivity and profitability of crops and cropping sequences. For instance, the productivity of maize increased 2 to 2.5 times under sole maize and fourfold under maize/pigeonpea intercropping system. Maize/pigeonpea intercropping system and maize-chickpea sequential system were identified as the most profitable ones. The area under maize, pigeonpea and maize-chickpea has increased more than three- and twofold, respectively.

Assessment of the economic benefits that have accrued due to the implementation of the watershed approach has revealed that the average net returns per hectare for dryland and irrigated cereals and pulses are higher within the watershed as compared to that of adjacent villages outside the watershed. Similarly, for pulse crops, per hectare net returns within the watershed are about twice as large as that outside the watershed. Implementation of holistic integrated watershed management has also resulted in increases in average household net income (Rs 15,400 within watershed as compared to Rs 12,700 outside watershed area). Compared to the 1998 levels, the evidence shows that farmer incomes in 2001 from crop production have doubled. Several factors have contributed to the impressive progress made in Adarsha watershed.

Mean annual rainfall at Kothapally watershed is about 826 mm with 50 rainy days and maximum rainfall being received during August and September. Computed annual potential evapotranspiration is about 1655 mm, which indicates atmospheric and crop water requirements are very high compared to rainfall. Climate change projections indicate to bring more volatility in rainfall with droughts and floods and pose additional challenges for sustainable crop production. General circulation model CESM1-CAM5 projects that the temperature in Kothapally is expected to increase in the range of 0.9 to 3.0 °C in the future under different scenarios. Similarly rainfall is projected to increase in the range of 37–120 mm. In general, climate change impacts positively on long-duration crops like cotton and pigeonpea due to increased rainfall in September and October, while negatively on short-duration crops like maize and sorghum. Under different scenarios and periods, cotton yield is expected to increase up to 10% and pigeonpea up to 6%. Similarly, maize and sorghum yield is projected to decrease by 5% and 6%, respectively. Integrated watershed management becomes more relevant for the sustainable crop yields under future climate change scenarios.

This case study has shown that with appropriate interventions and proactive participation of the beneficiary communities, watershed management can substantially improve the livelihoods of the poor in dryland areas while also enhancing the sustainability of resource use. Water conservation and access to improved germplasm have increased the profitability of otherwise unattractive conservation practices. Without access to improved varieties and markets, the conservation structures are unlikely to be attractive to individual farmers. The consortium approach to integrated watershed management has shown how the potential of marginal lands in predominantly rain-fed systems can be enhanced. We are sure that these results and lessons from Adarsha watershed will enhance the effectiveness of other watershed development programmes being undertaken by the government of India and in other countries.

References

Assefa, Y., Staggenborg, S. A., & Prasad, P. V. V. (2010). Grain sorghum water requirement and response to drought. *Crop Management.* https://doi.org/10.1094/CM-2010-1109-01-RV. (KAES: 10-296-J).

Basu, P. S., Ali, M., & Chaturvedi, S. K. (2009). Terminal heat stress adversely affects chickpea productivity in northern India – Strategies to improve thermo tolerance in the crop under climate change. *ISPRS Archives XXXVIII-8/W3 Workshop Proceedings: Impact of Climate Change on Agriculture,* pp. 189–193.

Baumgaertner, A. J. G., Jöckel, P., Kerkweg, A., Sander, R., & Tost, H. (2016). Implementation of the Community Earth System Model (CESM) version 1.2.1 as a new base model into version 2.50 of the MESSy framework. *Geoscientific Model Development, 9*, 125–135, 2016.

Bennett, A. J., Bending, G. D., Chandler, D., Hilton, S., & Mills, P. (2012). Meeting the demand for crop production: The challenge of yield decline in crops grown in short rotations. *Biological Reviews, 87*(1), 52–71.

Bhatia, V. S., Piara, S., Wani, S. P., Kesava Rao, A. V. R., & Srinivas, K. (2006). *Yield gap analysis of soybean, groundnut, pigeonpea and chickpea in India using simulation modeling.* Global theme on Agroeco systems report no. 31 (p. 156). Patancheru: International Crops Research Institute for the Semi-Arid Tropics (ICRISAT).

Birthal Prathap, S., Md. Tajuddin Khan, Digvijay S. Negi and Shaily Agarwal. (2014). Impact of climate change on yields of major food crops in India: Implications for food security. Agricultural Economics Research Review 27 (2) 2014 pp 145–155.

Byjesh, K., Naresh Kumar, S., & Aggarwal, P. K. (2010). Simulating impacts, potential adaptation and vulnerability of maize to climate change in India. *Mitigation and Adaptation Strategies for Global Change, 15*(5), 413–431.

Carberry, P. S., Hochman, Z., McCown, R. L., Dalgliesh, N. P., Foale, M. A., Poulton, P. L., Hargreaves, J. N. G., Hargreaves, D. M. G., Cawthray, S., Hillcoat, N., & Robertson, M. J. (2002). The FARMSCAPE approach to decision support: Farmers', advisers', researchers' monitoring, simulation, communication and performance evaluation. *Agricultural Systems, 74*, 141–177.

Climate Change Agriculture and Food Security (CCAFS). (2018). http://www.ccafs-climate.org/data_spatial_downscaling

DACNET. (2016). Latest APY State data, Directorate of Economics and Statistics, Government of India. http://eands.dacnet.nic.in/PDF/5-Year_foodgrain2010-15.xls

Deng, Z. W. (2010). Impact of climate warming and drying on food crops in northern China and the countermeasures. *Acta Ecologica Sinica, 30*(22), 6278–6288.

Descheemaeker, K., Oosting, S. J., Homann Kee-Tui, S., Masikati, P., Falconnier, G. N., & Giller, K. E. (2016). Climate change adaptation and mitigation in smallholder crop–livestock systems in sub-Saharan Africa: A call for integrated impact assessments. *Regional Environmental Change, 16*, 2331–2343.

Ericksen, P. J., Ingram, J. S. I., & Liverman, D. M. (2009). Food security and global environmental change: Emerging challenges. *Environmental Science & Policy, 12*, 373–377. https://doi.org/10.1016/j.envsci.2009.04.007.

FAOSTAT. (2018). Food and Agriculture Organization of the United Nations. http://fao.org/faostat/en/#data

Foley, A., Ramankutty, N., Brauman, K. A., Cassidy, E. S., Gerber, J. S., Johnston, M., Mueller, N. D., O'Connell, D., Ray, D. K., West, P. C., Balzer, C., Benett, E. M., Carpenter, S. R., Hill, J., Monfreda, C., Rockstrom, P., Sheehan, J., Siebert, S., Tilman, D., & Zaks, D. P. (2011). Solutions for a cultivated planet. *Nature, 478*, 337–342.

Gangadhar Rao, D., Katytal, J. C., Sinha, S. K., & Srinivas, K. (1995). Impacts of Climate Change on Sorghum productivity in India: Simulation study. American Society of Agronomy, 677 S. Segoe Rd. Madison, WI 53711, USA. *Climate change and Agriculture : Analysis of Potential International Impacts* (ASA Special publication No. 59).

Hagmann, J., Chuma, E., Murwira, K., & Connolly, M. (1999). *Putting process into practice: Operationalising participatory extension*. Overseas Development Institute.

Holzworth, D. P., Huth, N. I., deVoil, P. G., Zurcher, E. J., Herrmann, N. I., McLean, G., Chenu, K., et al. (2014). APSIM – Evolution towards a new generation of agricultural systems simulation. *Environmental Modelling & Software, 62*, 327–350. https://doi.org/10.1016/j.envsoft.2014.07.009.

Holling, C. S., & Meffe, G. K. (1996). Command and control and the pathology of natural resource management. *Conservation Biology, 10*(2), 328–337.

ICRISAT. (2002). *Giricidia for improving Soil Fertility*. Documentation. Patancheru, Andhra Pradesh, India: International Crops Research Institute for the Semi-Arid Tropics.

International Benchmark Sites Network for Agrotechnology Transfer. (1993). *The IBSNAT decade*. Honoluly: Department of Agronomy and Soil Science, College of Tropical Agriculture and Human Resources, University of Hawaii.

IPCC. (2014). Summary for policymakers. In C. B. Field, V. R. Barros, D. J. Dokken, K. J. Mach, M. D. Mastrandrea, T. E. Bilir, M. Chatterjee, K. L. Ebi, Y. O. Estrada, R. C. Genova, B. Girma, E. S. Kissel, A. N. Levy, S. MacCracken, P. R. Mastrandrea, & L. L. White (Eds.), *Climate change 2014: Impacts, adaptation, and Vulnerability. Part A: Global and sectoral aspects. Contribution of working group II to the fifth assessment report of the intergovernmental panel on climate change* (pp. 1–32). Cambridge: Cambridge University Press.

Jones, J. W., Tsuji, G. Y., Hoogenboom, G., Hunt, L. A., Thornton, P. K., Wilkens, P. W., Imamura, D. T., Bowen, W. T., & Singh, U. (1998). Decision support system for agrotechnology transfer; DSSAT v3. In G. Y. Tsuji, G. Hoogenboom, & P. K. Thornton (Eds.), *Understanding options for agricultural production* (pp. 157–177). Dordrecht: Kluwer Academic Publishers.

Jones, J. W., Hoogenboom, G., Porter, C. H., Boote, K. J., Batchelor, W. D., Hunt, L. A., et al. (2003). The DSSAT cropping system model. *European Journal of Agronomy, 18*(3–4), 235–265.

Kampen, J. (1982). An approach to improved productivity on deep Vertisols.

Karlen, D. L., Wollenhaupt, N. C., Erbach, D. C., Berry, E. C., Swan, J. B., Eash, N. S., & Jordahl, J. L. (1994). Crop residue effects on soil quality following 10-years of no-till corn. *Soil and Tillage Research, 31*(2–3), 149–167.

Kesava Rao, A.V. R., Wani, S. P., Srinivas, K., Singh, P., Bairagi, S. D., & Ramadevi, O. (2013). Assessing impacts of projected climate on pigeonpea crop at Gulbarga. *Journal of Agrometeorology*. Special Issue.

Liebman, M., Mohler, C. L., & Staver, C. P. (2001). *Ecological management of agricultural weeds*. Cambridge University Press.

Lipper, L., Thornton, P., Campbell, B. M., Baedeker, T., Braimoh, A., Bwalya, M., Caron, P., Cattaneo, A., Garrity, D., Henry, K., Hottle, R., Jackson, L., Jarvis, A., Kossam, F., Mann, W., McCarthy, N., Meybeck, A., Neufeldt, H., Remington, T., Sen, P. T., Sessa, R., Shula, R., Tibu, A., & Torquebiau, E. F. (2014). Climate-smart agriculture for food security. *Nature Climate Change, 4*, 1068–1072.

Liwenga, E. T. (2008). Adaptive livelihood strategies for coping with water scarcity in the drylands of Central Tanzania. *Physics and Chemistry of the Earth, 33*, 775–779.

Ludlow, M. M., & Muchow, R. C. (1990). A critical evaluation of traits for improved crop yields in water-limited environments. *Advances in Agronomy, 43*, 107–153.

Ludwig, F., & Asseng, S. (2006). Climate change impacts on wheat production in a Mediterranean environment in Western Australia. *Agricultural Systems, 90*, 159–179.

Luo, Q., Bellotti, W., Williams, M., & Bryan, B. (2005). Potential impact of climate change on wheat yield in South Australia. *Agricultural and Forest Meteorology, 132*, 273–285.

Ma, S., Zhang, S., Wang, F., Liu, Y., Liu, Y., et al. (2013). Highly efficient and specific genome editing in silkworm using custom TALENs. *PLoS One, 7*(9), e45035.

Mall, R. K., Lal, M., Bhatia, V. S., Rathore, L. S., & Singh, R. (2004). Centre for Systems Simulation, Indian Agricultural Research Institute, New Delhi 110012, India. *Agricultural and Forest Meteorology, 121*, 113–125.

Mishra, U., Jastrow, J. D., Matamala, R., Hugelius, G., Koven, C. D., Harden, J. W., Ping, C. L., Michaelson, G. J., Fan, Z., Miller, R. M., McGuire, A. D., Tarnocai, C., Kuhry, P., Riley, W. J., Schaefer, K., Schuur, E. A. G., Jorgenson, M. T., & Hinzman, L. D. (2013). Empirical estimates to reduce modeling uncertainties of soil organic carbon in permafrost regions: A review of recent progress and remaining challenges. *Environmental Research Letters, 8*, 035020. (9pp).

Murthy, M. V. R., Singh, P., Wani, S. P., Khairwal, I. S., & Srinivas, K. (2007). *Yield gap analysis of sorghum and pearl millet in India using simulation modeling* (Global Theme on Agroecosystems Report no. 37). International Crops Research Institute for the Semi-Arid Tropics, Patancheru 502324, Andhra Pradesh, India, 82 pp.

Naab, J., Bationo, A., Wafula, B. M., Traore, P. S., Zougmore, R., Outtara, M., Tabo, R., & Vlek, P. L. G. (2012). African perspective on climate change and agriculture: Impacts, adaptation and mitigation potential. In C. Rosenzweig & D. Hillel (Eds.), *Handbook of climate change and agroecosystems: Global and regional aspects and implications* (pp. 103–124). London: Imperial College Press.

National Food Security Mission Pulse Component. 2009. Annual Report. Department of Agriculture and Cooperation, Ministry of Agriculture, Govt. of India, New Delhi.

Pathak, P., Sahrawat, K. L., Wani, S. P., Sachan, R. C., & Sudi, R. (2009). Opportunities for water harvesting and supplemental irrigation for improving rainfed agriculture in semi-arid areas. In *Rainfed agriculture: Unlocking the potential* (pp. 197–221).

Rosegrant, M.W., Msangi, S., Ringler, C., Sulser, T.B., Zhu, T., Cline, S.A., (2008). International Model for Policy Analysis of Agricultural Commodities and Trade (IMPACT): Model description. Washington, DC: International Food Policy Research Institute. Accessed August 2010.

Sandeep, V. M., Bapuji Rao, B., Bharati, G., Rao, V. U. M., Pramod, V. P., Chowdary, P. S., Patel, N. R., & Vijayakumar, P. (2017). Projected future change in water requirement of grain sorghum in India. *Journal of Agrometeorol., 19*(3), 217–225.

Shantanu Kumar, D., Uma, S., & Singh, S. K. (2011). Impact of climate change on pulse productivity and adaptation strategies as practiced by the pulse growers of Bundelkhand region of Uttar Pradesh. *Journal of Food Legumes, 24*(3), 230–234.

Shiferaw, B., Anupama, G. V., Nageswara Rao, G. D. and Wani, S. P. (2002). *Socioeconomic characterization and analysis of resource-use patterns in community watersheds in semi-arid India* (Working Paper Series no. 12). Patancheru: International Crops Research Institute for the Semi-Arid Tropics, 44 pp.

Singh, P., Nedumaran, S., Traore, P. C. S., Boote, K. J., Singh, N. P., Srinivas, K., & Bantilan, M. C. S. (2014). Quantifying potential benefits of drought and heat tolerance in sorghum for adapting to climate change. *Agricultural and Forest Meteorology, 185*(2014), 37–48.

Sreedevi, T. K., & Wani, S. P. (2009). Integrated farm management practices and upscaling the impact for increased productivity of rain-fed systems. In S. P. Wani, J. Rockstorm, & T. Oweis (Eds.), *Rain-fed agriculture: Unlocking the potential* (Comprehensive assessment of Water Management in Agriculture Series) (pp. 222–257). Wallingford: CAB International.

Sreedevi, T. K., Shiferaw, B., & Wani, S. P. (2004). *Adarsha watershed in Kothapally: Understanding the drivers of higher impact* (Global theme on agroecosystems report no. 10). Patancheru: International Crops Research Institute for the Semi-Arid Tropics, 24 pp.

Tilman, D., Cassman, K. G., Matson, P. A., Naylor, R., & Polasky, S. (2002). Agricultural sustainability and intensive production practices. *Nature, 418*(6898), 671.

Tsuji, G. Y. (1998). Network management and information dissemination for agrotechnology transfer. In G. Y. Tsuji, G. Hoogenboom, & P. K. Thornton (Eds.), *Understanding options for agricultural production* (pp. 367–381). Dordrecht: Kluwer Academic Publishers.

Tsuji, G. Y., Uehara, G., & Balas, S. (1994). *Decision support system for Agrotechnology transfer (DSSAT) version 3*. Honolulu: University of Hawaii.

Uehara, G. (1989). Technology transfer in the tropics. *Outlook Agricultural, 18*, 38–42.

Uehara, G. (1998). Synthesis. In G. Y. Tsuji, G. Hoogenboom, & P. K. Thornton (Eds.), *Understanding options for agricultural production* (pp. 389–392). Dordrecht: Kluwer Academic Publishers.

United Nations, (2015). *World population prospects – Population Division.* https://esa.un.org/unpd/wpp/publications/files/key_findings_wpp_2015.pdf

Wani, S. P., Pathak, P., Tan, H. M., Ramakrishna, A., Singh, P., & Sreedevi, T. K. (2002). *Integrated watershed management for minimizing land degradation and sustaining productivity in Asia* (pp. 207–230). Integrated land management in dry areas: Proceedings of a Joint UNU-CAS International Workshop, 8±13 September 2001, Beijing, China (Zafar Adeel, ed.). Tokyo, Japan: United Nations University

Wani, S. P., Singh, H. P., Sreedevi, T. K., Pathak, P., Rego, T. J., Shiferaw, B., & Iyer, S. R. (2003). *Farmerparticipatory integrated watershed management: Adarsha watershed, Kothapally, India, an innovative and upscalable approach. A case study* (pp. 123–147). Research towards integrated natural resources management: Examples of research problems, approaches and partnerships in action in the CGIAR (Harwood, RR and Kassam AH, eds.). Washington, DC, USA: Interim Science Council, Consultative Group on International Agricultural Research; and Rome, Italy: Food and Agriculture Organization of the United Nations.

Wani, S. P., Ramakrishna, Y. S., Sreedevi, T. K. et al. (2006). Issues, concepts, approaches and practices in integrated watershed management: Experience and lessons from Asia. In: Integrated management of watersheds for agricultural diversification and sustainable livelihoods in eastern and Central Africa, 17–36. In *Proceedings of the international workshop held at ICRISAT, December 6–7, 2004, Nairobi, Kenya.* Patancheru: ICRISAT.

Wani, S. P., Joshi, P. K., Raju, K. V., Sreedevi, T. K., Wilson, M. J., Shah, A., Diwakar, P. G., Palanisami, K., Marimuthu, S., Jha, A. K., Ramakrishna, Y. S., Meenakshi, S. S., & D'Souza, M. (2008). *Community watershed as a growth engine for development of dryland areas. A comprehensive assessment of watershed programs in India, global theme on agroecosystems report no. 47* (pp. 1–46). Patancheru: International crops research Institute for the Semi-arid Tropics.

WorldClim 1.4 (current conditions) by www.worldclim.org; Hijmans et al. (2005). *International Journal of Climatology, 25*, 1965–1978. is licensed under a Creative Commons Attribution-ShareAlike 4.0 International License.

Chapter 7
Impacts of Integrated Watershed Development Using Economic Surplus Method

D. Moses Shyam, K. H. Anantha, S. P. Wani, and K. V. Raju

Abstract Adarsha watershed is a successful scientific narrative of sustainable integrated watershed programme conceptualized by ICRISAT team for efficient management of natural resources. Creating a proof of concept and a learning site for extension agents, NGOs, the national agricultural research system, policy makers and farmers was one of the main objectives of ICRISAT team when the institute started its work in the Adarsha watershed in Kothapally village, Ranga Reddy district, Telangana, India, in 1999. Water harvesting structures, 14 check dams, 97 gully control structures of loose stones, 1 gabion structure and others together have created a net storage capacity of 21,000 m³ which harvested nearly 70,000 m³ runoff water per year and have brought an additional area of 55 ha into irrigation by improving the groundwater table from 2.5 to 6.0 m. With improved technologies, farmers obtained high maize yields (28%) than the base year. Cotton has observed major yield gain (387%), major because of both technological change (Bt cotton) and assured water availability. Pigeon pea has recorded an increased productivity over the timeline (61%). Watershed has contributed to improve resilience of agricultural income despite the high incidence of drought in the watershed in 2002. Whilst drought-induced shocks reduced the average share of crop income in the non-watershed area from 44% to 12%, this share remained unchanged at about 36% in the watershed area. Livestock sector also contributed significantly to the total household income in watershed villages even during drought situations. Reduction in marginal cost due to supply shift has improved the cost-benefit ratio across the crops and ranged from 1.72 in cotton to 4.1 in pigeon pea. The BCR is worked out to be more than 2 and IRR

D. Moses Shyam (✉) · K. H. Anantha
ICRISAT Development Centre, International Crops Research Institute for the Semi-Arid Tropics (ICRISAT), Hyderabad, Telangana, India
e-mail: dmosesshyam@cgiar.org

S. P. Wani
Former Director, Research Program Asia and ICRISAT Development Centre, International Crops Research Institute for the Semi-Arid Tropics (ICRISAT), Hyderabad, Telangana, India

K. V. Raju
Former Theme Leader, Policy and Impact, Research Program-Asia, International Crops Research Institute for the Semi-Arid Tropics (ICRISAT), Hyderabad, Telangana, India

31%, implying that the returns to public investment such as watershed development activities were feasible and economically remunerative. The NPV worked out to be Rs. 141 lakh INR for the entire watershed. The total treated area in the watershed was around 465 ha, and the NPV per ha worked out to be Rs. 30,000 INR which implied that the benefits from watershed development were higher than the cost of investment of the watershed development programs. The study revealed that the watershed development has the potential for poverty reduction by generating impressive returns on investment even during drought year. The new generation watershed intervention emphasizes achieving the food and income security of farmers while maintaining the integrity of the eco-hydrology and other natural systems in the watershed.

Keywords Watershed · Impact · Economic surplus · Kothapally

7.1 Background

Recognizing the importance of dryland agriculture for individual as well as national food security and environmental security, the government of India, NGOs and multilateral donor agencies have, for decades, developed and promoted innovations that aim to increase the value and productivity of drylands through several development programs with different approaches. Watershed development program is one such integrated holistic approach dealing with multidisciplinary issues for sustainable development in these fragile areas (Kerr et al. 2000; Wani et al. 2003). Watershed approaches have evolved from externally imposed biophysical interventions towards more participatory approaches, encompassing a broader range of activities that have a potential impact on holistic livelihood activities. In many instances, such innovations not only attempt to increase productivity but also mitigate the climatically induced uncertainty of production through specific soil, crop and rainfall management strategies (Wani et al. 2003, 2008, 2011a, b; Wani and Ramakrishna 2005). Such research has already shown great potential on research stations and in farmers' fields, with achievable yields often several times greater than those obtained by traditional farmers' practice (Wani et al. 2003; Sreedevi et al. 2004, 2006; Pathak et al. 2007; Wani and Johan 2011). In addition, watershed development interventions are useful in restoring and rejuvenating several ecosystem services at a microscale through proper land use planning (Garg et al. 2011).

In India, integrated watershed programs have silently revolutionized the developments in drylands (Wani et al. 2009). There are few watersheds which are performing well in India because of innovative approaches (technical, social, institutional and linkages). According to Palanisami and Kumar (2009), watershed development activities have been found to increase crop productivity, crop intensity and diversification leading to increased employment. Wani et al. (2009) state that focusing more on integrated watershed management plays an important role in ensuring food security, reducing poverty, protecting the environment and addressing issues such as equity and improved livelihoods.

Kothapally watershed in Telangana state in India is one of the communities ICRISAT team worked closely with starting in 1999. Key to the success has been ensuring the community is empowered to drive the innovations with ICRISAT taking a catalyst role and providing the scientific backing to all the interventions. Over the decades, the village has prospered and a holistic approach is developed as more innovations were introduced. These started with water and soil management and improved crop varieties and diversity on farm and later expanded to include livestock integration, linking farmers to markets and building alternative livelihoods, wastewater treatment, self-sustaining filtered drinking water, and more. These measures were implemented on common resources, viz. water courses, *nala* and wastelands. The committee members had identified 21 potential sites for water storage structures (small check dams), 270 sites for gully control structures, 11 gabion structures, 38 ha for field bunding, and a 500 m long diversion bund to avoid damage to crop lands. Fourteen water storage structures (one earthen and 13 masonry) with a capacity of 300 – 2000 m³ water storage were constructed. Ninety-seven gully control structures, 60 mini percolation pits, one gabion structure for increasing groundwater recharge, a 500 m long diversion bund and field bunding on 38 ha were completed. Twenty-eight dry open wells, near nala (small streams), were recharged through runoff water flowing in the *nala* during runoff events. A users' group was formed for each water storage structure, and the water collected in the storage structures was exclusively used for recharging the groundwater as resolved and decided by the Watershed Committee.

In this context, there is a need to identify the potential drivers of success and also to understand the shortcomings that may affect the sustainability of the benefits. These findings and lessons will be useful not only for Kothapally watershed but also to other dryland watershed communities now grappling with improving livelihoods of the people and protecting the productive resource base.

7.2 A Glimpse of Watershed Development Programs in India

Watershed development programs have been initiated in India since 1970 with emphasis on rainwater harvesting and subsequently upgraded to soil and water conservation, to improve the productivity of dryland which are drought-prone regions also through increased investments (Wani et al. 2008). The primary aim of the watershed development programs is to increase the production potential of dryland regions and to meet the needs of rural communities on sustainable basis for food, fuel, fodder, etc., thereby reducing the pressure on existing production zones. Further, the objective of the program has undergone substantial modifications to include and address varying components of rural livelihoods. Thus, the watershed programs have evolved from being purely technically oriented soil-water conservation programs to more integrated and participatory approach aiming at natural resource management with organization of beneficiaries (GoI 2001a, b, 2007) and more recently targeting holistic livelihood improvement (GoI 2008). Therefore, the approach shifted from traditional top-down approach to more holistic participatory approach to address sustainability and transparency through community participa-

Table 7.1 Summary of benefits from the selected watershed studies in India

	Particulars	Unit	No. of studies	Mean	Min	Max	t-value
Efficiency	B:C ratio	Ratio	311	2.01	0.82	7.30	35.09
	IRR	Per cent	162	27.43	2.03	102.70	21.75
Equity	Employment	Person Days/ha/year	99	154.53	0.05	900.00	8.13
Sustainability	Increase in irrigated area	Per cent	93	51.55	1.28	204.00	10.94
	Increase in cropping intensity	Per cent	339	35.51	3.00	283.00	14.96
	Runoff reduced	Per cent	83	45.72	0.38	96.00	9.36
	Soil loss saved	t/ha/year	72	1.12	0.11	2.05	47.21

Source: Joshi et al. 2008

tion (Wani et al. 2006). Evidence from a cross-section of watersheds for India during the last four decades suggests that watershed development programs have yielded significant economic and environmental benefits (Kerr et al. 2000; Deshpande and Narayanamoorthy 1999; Wani et al. 2003; Chandrakanth et al. 2004; Wani et al. 2008, 2011a, b; Garg et al. 2011). A meta-analysis of 633 case studies, presented in Table 7.1, reconfirmed the fact that watershed projects yielded multiple exemplary benefits in terms of economics, environment and equity parameters in varied climatic conditions (Joshi et al. 2008). Further, it revealed that the benefits from watershed programs were conspicuously more in the low-income regions as compared with the high-income regions (Joshi et al. 2008).

The watershed development programs are receiving increasing support extended by a number of international funding agencies along with support from national and state governments. These agencies also sponsor and implement watershed development projects, but a significant proportion of the investment in these projects is being made by the Government of India. Since the 1970s, the Government of India has allocated substantial amount of resources ($7 billion) for improving dryland areas through watershed development programmes (Joshi et al. 2005). Evidence from a cross-section of watersheds for India during recent years suggest that watershed development programmes have yielded significant economic and environmental benefits (Joshi et al. 2008). Given the focus of the federal government on using watershed programme as an important tool in accelerating development of dryland regions of the semi-arid tropics, it becomes imperative to assess the impact of these programs on agriculture sector.

7.3   Methodology

7.3.1   Study Area

The data used for this study is collected from Adarsha watershed of Kothapally in Telangana, a semi-arid watershed located between longitudes 78.27° east and latitudes 17.53° north and at an elevation of 500 m above mean sea level (Fig. 7.1). The

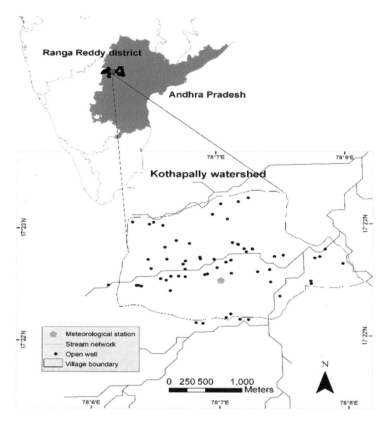

Fig. 7.1 Location map of Adarsha watershed, Kothapally, Telangana, (erstwhile Andhra Pradesh), India

watershed receives an average of 750 mm rainfall during the monsoon season, starting from June to October. However, rainfall is highly erratic both in terms of total, intensity and distribution over time. Kothapally is a village of nearly 266 households depending largely on agriculture, either as owner-cultivators or landless labourers. About 70% of the farmers are smallholders having less than 2 ha of land. Within the village boundaries, there are 62 open wells, most of which occur along the main watercourse. These open wells are limited in depth, a typical well depth being between 15 and 35 ft. There were 15 bore wells before watershed project initiation, and 55 new bore wells were dug during the project. What is very remarkable is that, in Kothapally, there are no deep tube wells. In 1999, watershed project was implemented in the village, and it covers about 465 ha and has medium to shallow black soils, with a depth of 30–90 cm (Wani and Shiferaw 2006).

Watershed activities were undertaken in the village during the period 1999–2004, with the help from the Government of Andhra Pradesh, Asian Development Bank and the consortium partners. Before the commencement of the watershed development activities, agricultural activity in the watershed was limited to the rainy season only with monocropping system with low yield. The Kothapally watershed was selected for agricultural water interventions for several reasons: (1) more than 90% of the cultivable area was rainfed, characterized by water scarcity; (2) crop productivity was below 0.5–1.0 ton/ha; (3) many open wells were defunct, and the community experienced acute water shortage for drinking purposes, especially during the summer period; and (4) the non-existence of water harvesting structures and the potential for minimum interventions to conserve soil and water (Wani et al. 2003).

7.3.2   Economic Surplus Approach

The Economic Surplus (ES) approach is widely followed for evaluating the impact of technology on the economic welfare of households (Moore et al. 2000; Wander et al. 2004; Maredia et al. 2000; Swinton 2002; Palanisami et al. 2009). With this model, aggregate social benefits (consumer and producer surplus) of a project within a target domain were estimated for any technological change originated by research.

7.3.3   Theoretical Framework

The model is based on the Marshallian theory of economic surplus that stems from shifts over time of the supply and demand curves. The supply shift (S_0) due to changes in production technology (S_1) creates economic surplus to both producer and consumer (Fig. 7.2). With the shift in supply, the original market equilibrium a (P_0, Q_0) is transferred by the effect of technological change to b (P_1, Q_1).

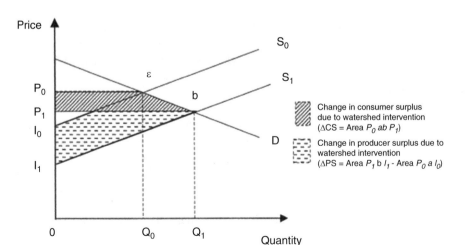

Fig. 7.2 Graphical representation of economic surplus method (Source: Palanisami et al. 2009)

Consumers gain because they are able to consume a greater amount (Q_1) at a lower price (P_1). The area P_0abP_1 represents the consumer surplus. The watershed development intervention affects agricultural producers in two ways: (1) lower marginal costs (according to the theory, the supply curve corresponds to the curve of marginal costs as of the minimum value of the curve of average variable costs) and (2) lower market price (P_0 reduced to P_1). Thus, the producers' surplus is defined as the Area P_1bl_1 − Area P_0al_0.

The mathematical model used was based on the scheme proposed by Pachico et al. (1987), in which supply and demand functions were nonlinear with constant elasticity, i.e. log-linear. The supply function for a product market was assumed that supply curves of the following functional form:

$$s_0 = c\left(P_0 - P_{10}\right)^d \tag{7.1}$$

where

s_0 = initial supply before watershed intervention,
c, d = constants, P_0 = price of product and
P_{lo} = minimum price that producers are willing to offer.

Due to technological interventions, the output supply curve shifts gradually over time benefiting the agricultural sector through water resource enhancement (Palaniswamy et al. 2009). This supply shift factor due to technological change is known as K. The supply shift factor (K) can be interpreted as a reduction of absolute costs for each production level or as an increase in production for each price level (Libardo et al. 1999).

Microeconomic theory defines consumer surplus (individual or aggregated) as the area under the (individual or aggregated) demand curve and above a horizontal line at the actual price (in the aggregated case: the equilibrium price). Following IEG, World Bank, 2008, the demand curve is assumed to be log-linear with constant elasticity. Thus, the demand equation for this demand function can be written as:

$$P = gQ^\eta \tag{7.2}$$

where η is the elasticity and g is a constant. Once, the parameters η and g are estimated, and then consumer surplus could be estimated by Eq. (7.3):

$$CS = \int_{Q_0}^{Q_1} gQ^\eta \, dQ - \left(Q_1 - Q_0\right)P_1 \ldots \tag{7.3}$$

Combined, the consumer surplus and the producer surplus make up the total surplus.

7.3.4 Estimation of Benefits

Following the theory of demand and supply equilibrium, economic surplus (benefits) as a result of watershed development intervention is measured by Eq. (7.4):

$$B = K * P_0 * A_0 * Y_0 * \left(1 + 0.5 Z * \varepsilon_d\right) \qquad (7.4)$$

where K is the supply shift due to watershed intervention.

The supply shift due to watershed intervention can be mathematically represented by Eq. (7.5):

$$K = V * \rho * \Psi * \Omega \qquad (7.5)$$

where K represents the vertical shift of supply due to intervention of watershed development technologies and is expressed as a proportion of initial price. V is the net cost change which is defined as the difference between reduction in marginal cost and reduction in unit cost. The reduction in marginal cost is defined as the ratio of relative change in yield to price elasticity of supply (ε_s). Reduction in unit cost is defined as the ratio of change in cost of inputs per hectare to (1 + change in yield). ρ is the probability of success in watershed development implementation. Ψ represents adoption rate of technologies and Ω is the depreciation rate of technologies.

Z represents the change in price due to watershed interventions. Mathematically, Z can be defined by Eq. (7.6):

$$Z = K * \frac{\varepsilon_s}{\left(\varepsilon_d + \varepsilon_s\right)} \qquad (7.6)$$

where P_0, A_0 and Y_0 represent prices of output, area and yield of different crops in the watershed before implementation of watershed development programme. If we use this with and without approach, then these represent area, yield and price of crops in control village.

7.3.5 Cost of Project

The analysis considered cost towards watershed development investment during the project period and maintenance expenditure incurred in the project. For watershed development projects with multiple technologies or crops, incremental benefits from each technology and crop were added to compile the total benefits. The worthiness of the watershed development projects was then evaluated at 10% discount rate. Using above estimates of returns and costs, net present value (NPV), benefit-cost ratio (BCR) and internal rate of return (IRR) were computed.

7.4   Results and Discussion

Watershed creates both short-term and long-term impacts (Govind[1] 2004). Short-term impacts include crop diversification/intensification and changes in crop, livestock and employment productivity. Long-term impacts include area expansion and environmental resilience. Adarsha watershed is best suited example to showcase the strength of science-led consortium which transformed the lives of rural people through building resilience as well as improving livelihoods through sustainable use of natural resources. Watershed intervention has made substantial impacts on the livelihoods of the people in the village. Supplementary irrigation has brought diversification of income and reduced vulnerability to drought and other stresses. The watershed has shown great impact on people's livelihood which is clear from the results presented below.

7.4.1   Land Holding Categories

There have been considerable changes in landholding size due to watershed interventions, and the results are presented in Table 7.2. The watershed area consisted of 270 households in year 1999, and this was marginally declined by 1.5% in year 2017. The change was observed in marginal and small farmer category where marginal farmers in the village declined by 52% and small farmers increased by 179%. Similar is the trend between farmers between medium and large farmers category. Finally, it can be concluded that the watershed interventions has resulted in a shift from marginal and medium farmers to small farmers.

Table 7.2 Distribution of households based on size of land holding

Land holding (ha)	1999	2017	% change
<1	135 (50)	65 (24)	−52
1–5	59 (22)	166 (63)	179
5.1–10	73 (27)	26 (10)	−64
>10	3 (1)	9 (3)	233
Total agricultural HH	270	266	−1.5

Source: Primary Survey

---

[1] Govind Babu, R.K Singh; Babu Singh,2004; Socio-economic impact of watershed development in Kanpur, Agricultural Economics Research Review, Vol. 17 (Conference No.) pp. 125–130.

7.4.2   Water Harvesting and Groundwater Recharge

Water harvesting structures together have created a net storage capacity of 21,000 m^3 which harvested nearly 70,000 m^3 runoff water per year facilitating groundwater recharge. Groundwater table has increased from 2.5 to 6.0 m after the interventions. About 30–40 ha field bunding has been done in farmers' fields to conserve rainwater as well as to reduce soil erosion. Runoff water was diverted into dry open wells to rejuvenate them as well as to increase rainwater use efficiency. A study showed that nearly 60% of the runoff water was harvested through agricultural water management interventions which also recharged shallow aquifers. Water harvesting structures resulted in a total 6 m rise in the water table during the monsoon. At the field scale, water harvesting structures recharged open wells at a 200–400 m spatial scale (Garg and Wani 2012). In addition to these, two decentralized domestic wastewater treatment units have been constructed, which benefits six farmers to use the treated water for crops like cotton, maize and vegetables while addressing the sanitation issue of the village that benefits the community in the village.

7.4.3   Historical Changes in Irrigated Area and Water Utilization Pattern

Water impounded per year has brought a significant change in groundwater availability in the village which is evident from the Fig. 7.3. Historical changes in area brought under irrigation were tracked, and a total of 110 ha of land was found to be brought under irrigation from 1950 to 2017. The figure shows that till watershed interventions, i.e. 1999, the area under irrigation was only 44 ha, and from 2000 to 2005 (till completion of watershed structures), another 10.5 ha of land was brought under irrigation. However, after completion of structures, there was an unprecedented growth in irrigation area expansion which brought an additional 55 ha of land in span of 12 years, i.e. from 2006 to 2017.

The annual water utilization by major crops (paddy, cotton, maize and vegetables) is presented in Fig. 7.4. The analysis revealed that the initial phase of watershed development has seen low water utilization due to limited scope as water was the scarce resource in the watershed. However, as watershed development progressed with construction of water harvesting structures and adoption of in situ water management practices, the cropping pattern has shifted from low water consuming crops to water intensive crops such as rice, vegetables and fodder grasses. Therefore, the water utilization pattern has shifted drastically after 2004 when construction of water harvesting structure was completed and the structures facilitated groundwater recharge and helped farmers to pump water for irrigation. From 2005 to 2014, about 80,000–85,000 m^3 water was utilized per annum by major crops in the watershed (Fig. 7.4). This indicates the potential opportunities for extending irrigated area through cost-effective water harvesting structures in the semi-arid tropics.

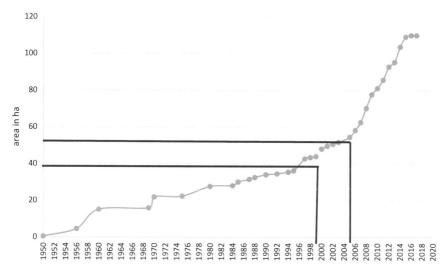

Fig. 7.3 Change in area under irrigation in Adarsha watershed, Kothapally

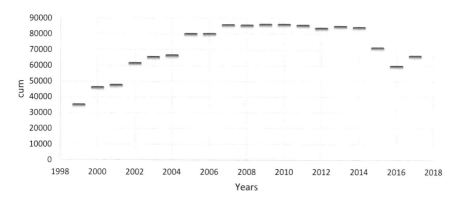

Fig. 7.4 Annual water utilization pattern by major crops in Adarsha watershed, Kothapally

7.4.4 Crop Mapping

The field inventory conducted at individual crop plots level was compiled systematically for understanding the agricultural holding pattern in Kothapally. The standard land holding classification (PIB 2015) has been adopted. The land holding pattern in Kothapally is shown in Figs. 7.5 and 7.6. The IRS-IC and ID LISS-III images in April 2000 and the NDVI images generated revealed an increase in vegetation cover from 321 ha in 1999 to 200 ha in 2000 and to 362 ha in 2017. However, this change is not continuous, and the land vegetation has reached its maximum by 2013 (478 ha) occupying nearly 95% of the total geographical area and later on starts declining due to various socio-economic reasons.

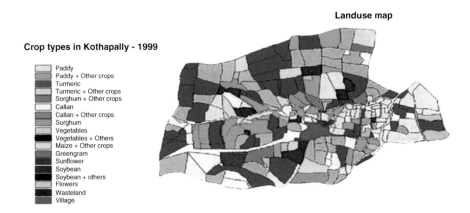

Fig. 7.5 Land use pattern in year 1999

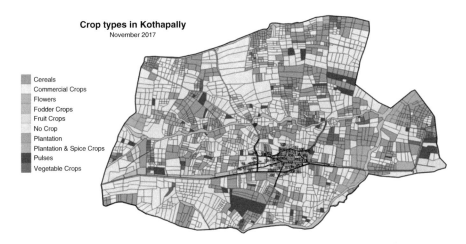

Fig. 7.6 Land use pattern in year 2017

7.4.5 Changes in Cropping Pattern

Kothapally was predominantly a cotton-growing area prior to project implementation (Fig. 7.7). The area under cotton was 200 ha in 1998. Maize, chickpea, sorghum, pigeon pea, vegetables and rice were also grown. After 19 years of activities in Adarsha watershed, the area under cotton cultivation marginally declined by 12% with simultaneous increases in maize. However, this decline in area under the crops is not uniform, and the cropped area has shown a positive trend till 2013 and started declining thereafter. Changes in rainfall, increased input costs and not getting the remunerative prices might be the possible reasons for the shifts in cropping pattern.

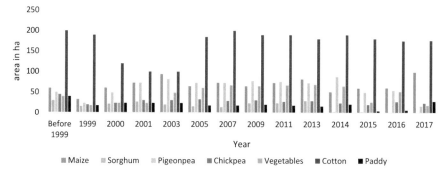

Fig. 7.7 Cropping pattern in Adarsha watershed, Kothapally, from 1999 to 2017

7.4.6 Changes in Productivity and Crop Income Levels

Increased Productivities Farmers evaluated improved crop management practices (INM, IPM and soil and water management) together with researchers. With improved technologies, farmers obtained high maize yields (28%) than the base year (Table 7.3). Cotton has observed major yield gain (387%), major because of both technological change (Bt cotton) and assured water availability. The yield levels of Sorghum have shown a declining trend might be the reason for declining area under the crop in the region. Pigeon pa has been recorded an increased productivity over the timeline (61%).

Reduction in marginal cost due to supply shift has improved the cost-benefit ratio across the crops and ranged from 1.72 in cotton to 4.1 in pigeon pea. Fluctuation in CB-ratio of cotton can be attributed to changes in farm gate prices.

7.4.7 Household Income

It is evident that watershed helps the farmers to build resilience during drought year in terms of sustaining crop income which is major indicator of food security. We can observe a clear system of diversification in watershed and non-watershed areas as nearly 50% of household income was derived from outside farm economy. The contribution of non-farm income is even higher (nearly one-third share of total income) in non-watershed areas during drought year as there is little scope for farm activities like in watershed areas. The crop income which is a major source of food security in rural areas contributes less than 50% both in watershed and non-watershed areas. However, watershed has contributed to the improved resilience of agricultural income despite the high incidence of drought during 2002 in the watershed. Whilst drought-induced shocks reduced the average share of crop income in the non-watershed area from 44% to 12%, this share remained unchanged at about

Table 7.3 Changes in crop productivity and income

Crops	Year	Grain yield kg ha⁻¹	Cost of cultivation Rs ha⁻¹	Total income Rs ha⁻¹	Profit Rs ha⁻¹	C:B ratio
Maize	1999	3250	4050	11,375	7325	1.81
	2000	3756	4038	14,271	10,234	2.53
	2001	3300	4265	13,530	9265	2.17
	2002	3481	4432	14,620	10,188	2.30
	2003	3920	5185	17,640	12,455	2.40
	2004	3421	5316	17,106	11,790	2.22
	2005	3920	6853	19,992	13,139	1.92
	2006	3635	7721	22,718	14,997	1.94
	2007	4680	8590	25,445	16,855	2.96
	2008	4808	9458	28,172	18,713	2.98
	2009	3830	10,327	30,899	20,572	2.99
	2017	4165	31,850	70,805	38,955	2.22
% change in yield	28	Average C:B ratio	2.37			
Pigeon pea	1999	643	2100	9006	6906	3.29
	2000	935	2442	14,026	11,584	4.74
	2001	799	2504	11,584	7762	3.10
	2002	719	2638	10,788	8149	3.09
	2003	949	2833	14,232	11,399	4.02
	2004	680	2945	12,234	9289	3.15
	2005	924	3112	16,640	13,529	4.35
	2006	970	3518	19,406	15,889	4.52
	2007	640	3924	22,172	18,249	4.65
	2008	760	4330	24,938	20,609	4.76
	2009	830	4736	27,704	22,969	4.85
	2017	1041	11,025	52,063	41,038	4.72
% change in yield	61.9	Average C:B ratio	4.10			
Sorghum	1999	3051	4003	13,728	9726	2.43
	2000	3171	3886	17,442	13,556	3.49
	2001	2600	4554	17,307	12,753	2.80
	2002	2425	4768	16,975	12,207	2.56
	2003	2288	4950	18,302	13,352	2.70
	2004	2324	5325	20,918	15,593	2.93
	2005	2250	5713	21,375	15,663	2.74
	2006	2086	6775	23,861	17,086	2.52
	2007	–				
	2008	–				
	2009	–				
	2017	2327	17,150	81,462	64,312	4.75

(continued)

Table 7.3 (continued)

Crops	Year	Grain yield kg ha⁻¹	Cost of cultivation Rs ha⁻¹	Total income Rs ha⁻¹	Profit Rs ha⁻¹	C:B ratio
% change in yield	−23.73	Average C:B ratio	2.99			
Cotton	1999	402	5345	8033	2688	0.50
	2000	1164	16,993	26,779	9786	1.58
	2001	1201	17,380	26,418	9038	1.52
	2002	1238	17,033	27,243	10,210	1.60
	2003	1303	19,175	31,260	12,085	1.63
	2004	1235	15,268	25,947	10,679	2
	2005	1797	15,794	34,141	18,347	2.16
	2006	2091	16,932	36,338	19,405	2.15
	2007	–				
	2008	–				
	2009	–				
	2017	1960	42,982	99,960	19,406	2.33
% change in yield	387.56	Average C:B ratio	1.72			

36% in the watershed area. Livestock sector also contributed significantly to total household income in watershed villages even during drought situations (Fig. 7.8).

Anantha and Wani (2016) studied the returns to land for major crops in the watershed. The study revealed that during the watershed intervention period, farmers with different crops realized returns to land amounting to US $720 ha⁻¹ from cotton followed by $295 ha⁻¹ from flowers, $287 ha⁻¹ from vegetables and $171 ha⁻¹ from cereals. However, the returns to land seem to be on declining trend during post-project intervention. Nevertheless, such returns to land do not vary much from each other because during the year with normal rainfall, the runoff is able to reach the end plots and subsidizes all other requirements. The overall average returns to land of $234 ha⁻¹ was realized during watershed intervention period, and it was same even during post-project period. Such a level of 'return to land' realized is substantial in the context of rural economy. This implies that the watershed intervention has built resilience for adapting to changing climate risks (Anantha and Wani 2016).

7.4.8 Assets

In 1998–1999, Kothapally village was less developed and there are no transport facilities. Eighty per cent of its 462 ha of agricultural land was rainfed, growing one crop per year. With the help of district administration, a number of programs were brought into the village. During 1999, watershed development program has been implemented under drought-prone area program. In terms of assets, with less than ten

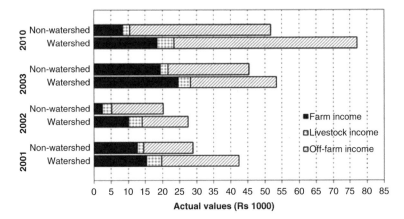

Fig. 7.8 Impact of watershed development approach on household income in Adarsha watershed, Kothapally. (Source: Anantha and Wani 2016)

autos, two luggage vans (transport vehicles), four lorries and nine tractors, at present, the village is buzzing with activities, and the number of autos grew up to 35; cars, 7; bikes, 174; tractors, 62; and lorries, 15. This indicates increase in economic activities brought by the integrated watershed interventions coupled with other rural development programs. The village has transformed into a fully developed and economically active village with majority of households converted their house into a *pucca (concrete)* house and availed drinking water and sanitation facility.

7.4.9   International Visitors Comments

Visit to Kothapally Watershed
by Emma on January 12, 2018
    Today, both the Ag systems team and rural infrastructure team visited the Adarsha watershed, in Kothapally. This development has allowed the village access to drinking and irrigation water year-round, which was not possible before. Each small water harvesting structures benefits two to three farmers and is relatively cheap to construct (2000–3000 rupees or $30–$50). Access to these structures has allowed the village to irrigate its fields throughout the year. As a result, farmers have been able to have as many as three crops per year, instead of only one. Added crops are often of higher value, like vegetables and fresh cut flowers. Overall yields have also increased 25–85%. It is clear to see then how incomes for these farmers have, in some cases, quadrupled. The farming system of the village has also witnessed drastic changes

(continued)

from cotton that had been widely grown by farmers, but with access to reliable irrigation, some 18–20 cropping systems now exist.

Another interesting facet of Adarsha is that it acts as an example and learning centre for villages and organizations interested in creating their own water management areas. Being a successful intervention, it has the potential of influencing further watershed developments across India. The success of this watershed, we gathered, was due to the acceptance of Kothapally's farmers and community. Their cooperation with facilitators like the government, and later ICRISAT, has ultimately led to the success of the project. As farmers benefited from the access to water, a trickle-down effect was seen throughout the village, improving the livelihoods of both women and children. I think this highlights the benefits of a watershed at the community level and reveals the social changes that can accompany agricultural development. Visiting Kothapally was therefore a rewarding experience and a true testament to the accomplishments of rural and agricultural development.

Key Takeaways Are

- Kothapally is a village in Telangana state.
- ICRISAT (CGIAR) was giving technical backup to the village with scientific planning, social engineering and Topographic survey, and water balance model were involved in implementing the programme.
- Constructed 16 check dams, 45 percolation tanks some sunken pits.
- Uniqueness of watershed: social entrepreneurship farming to non-farming sectors (45 SHGs).
- Groundwater recharge increased to 32% (now no scarcity of water for drinking and also to agriculture).
- Increased income of people by 50%.
- Cropping intensity increased 100–300%.
- Decreased water runoff/overflow to the tune of 5–7%.

Source: https://blogs.cornell.edu/internationalag6020/2018/01/12/visit-to-kothapally-watershed/

The Adarsha Alliance

Frank A Hilario – 12th September 2010

The impact of the success story of Adarsha has been worldwide, not only in India, where it is based, but also in Vietnam, Thailand, Africa, China and the Philippines. *The Adarsha Alliance*, the concept inspired by the success of the Adarsha villagers in India in waging a grey-to-green revolution within their village by first restoring their watershed. The Adarsha Alliance has been able to grow a watershed where none grew before.

Author opined 3 crucial points that I believe have made Adarsha a success story unprecedented in the history of science for development, in a direct sense a people power triumph.

1. **Common eye.** As it has turned out, as I see it, *Adarsha is the triumph of people commonly at work for a goal after they finally commonly have seen it as one for all*. I know quite a few short-lived success stories of people empowerment projects in the Philippines, where I am based. They all mean well, but good intentions are not good enough; where they have failed is that they are project-driven until project funds run out on all of them. In basketball, you can be a great team, but if you can't see the goal, you can only shoot wildly until you can't shoot anymore.
2. **Collective mission.** The mission is *to help the poor help themselves.* A basic rule, not to be ignored. The villagers of any Adarsha candidate in any place in the world will need to learn to walk the talk of people power, *real* people power. The end is the same: self-reliance. You want it done, right? Do it yourself!
3. **Collaboration of 3.** The triad is the scientists, villagers and their partners in grime. They all have to learn to work the dirt together. Now, that's not an enumeration by hierarchy, the more important people listed first; it is merely a listing by chronology, because it is usually the scientists who travel for troubles, looking for people's problems. Part of people power should be to arouse people's initiative from inside and among themselves. In the ICRISAT experience, the collaborators are the Indian government, other science institutions, private sector and donors such as the World Bank and Bill Gates, knowingly sharing for the poor people working to enrich themselves.

This describes what I shall now call *The Adarsha Team, the ideal group at work.* This I believe is one of the biggest contributions of ICRISAT to the universe of science in the service of the world of the poorest of the poor.

Source: https://icrisatwatch.blogspot.com/2011/07/?view=classic

Table 7.4 Cost of rainwater harvesting structures

Sl. no.	Structure type	No. of WHS	Unit cost ('000 Rs)	Total cost ('000 Rs)
1	Boulder check/gully plugs	97	5	485
2	Masonry check dam	14	80	1120
3	Gabion structure	1	25	25
4	Earthen check dam	2	50	100
	Total			1730

7.4.10   Cost of Structures

The total cost of all soil and water conservation structures was INR 17.30 lakhs which included 14 check dams (INR 11.20 lakhs), 97 gully control structures of loose stones (INR 4.85 lakhs) and 1 gabion structure (INR 0.25 lakhs). The cost details for other structures are presented in Table 7.4.

7.4.11   Research and Structure Costs

The annual flow of research costs and research benefits provides a deeper under-standing about welfare gains due to watershed interventions. The research and structure costs including the extension costs of ICRISAT scientific staff were con-sidered from 2000 to 2018 for calculation of project costs (Table 7.5). The flow of costs **was discounted with 10% discount rate** for the project period. The resulting net present value (NPV) was calculated by taking the differences between total dis-counted costs and discounted research benefits.

Tables 7.6 and 7.7 describe the flow of investment made on water harvesting structures, their maintenance and research and development interventions by the implementing agency and benefits accrued from farm activities. While working out the cost, all the costs including the costs on watershed treatment activities were included. For illustration purpose, the cost streams are assumed to be constant dur-ing the project period. The cost on water harvesting structures was ranging between 0.5 and 2.83 lakhs. However, maintenance and research costs are the major costs made in the watershed totalling about 200 lakhs.

7.4.12   Economic Surplus

The change in total surplus due to watershed development activities was estimated and has been presented in Table 7.7. The change in total surplus was higher in cot-ton, rice, pigeon pea and vegetables than crops like maize, chickpea and sorghum. Being the major cash crops, these crops have benefited more from the watershed

Table 7.5 Flow of structure and research costs

Year	Structure cost (actuals-in Lakh Rs)	Maintenance cost (actuals-in Lakh Rs)	Research cost actuals-in Lakh Rs)	Total cost (Lakh Rs)
1999	0.55		14.87	15.42
2000	1.80		14.87	16.67
2001	0.93		14.87	15.8
2002	3.36		14.87	18.23
2003	2.58		14.87	17.45
2004			14.87	14.87
2005	0.64	6.745	14.87	22.255
2006			6.887	6.887
2007			6.887	6.887
2008			6.887	6.887
2009		6.745	6.887	13.632
2010			6.887	6.887
2011			6.6626	6.6626
2012			6.6626	6.6626
2013		6.745	6.6626	13.4076
2014			6.6626	6.6626
2015			5.8626	5.8626
2016			5.8626	5.8626
2017		6.745	5.8626	12.6076
Total	8.7	26.98	**182.7632**	**219.603**

Table 7.6 Watershed cash flow statement in Adarsha watershed, Kothapally

Year	Costs	Benefits	Cash flow	Discount factor	Discounted benefits	Discounted costs
2000	14.87	3.5	−11.4	0.9091	3.1	13.5
2001	14.87	7.4	−7.5	0.8264	6.1	12.3
2002	14.87	7.1	−7.7	0.7513	5.4	11.2
2003	14.87	8.8	−6.1	0.6830	6.0	10.2
2004	14.87	11.1	−3.8	0.6209	6.9	9.2
2005	21.62	98.4	76.8	0.56447	55.6	12.2
2006	6.89	130.7	123.8	0.51316	67.1	3.5
2007	6.89	149.4	142.5	0.46651	69.7	3.2
2008	6.89	161.5	154.6	0.42410	68.5	2.9
2009	13.63	186.0	172.3	0.38554	71.7	5.3
2010	6.89	199.9	193.0	0.35049	70.1	2.4
2011	6.66	223.6	216.9	0.31863	71.2	2.1
2012	6.66	241.9	235.2	0.28966	70.1	1.9
2013	13.41	265.7	252.3	0.26333	70.0	3.5
2014	6.66	215.2	208.5	0.23939	51.5	1.6
2015	5.86	172.4	166.5	0.21763	37.5	1.3
2016	5.86	220.4	214.5	0.19784	43.6	1.2
2017	5.86	204.7	198.9	0.17986	36.8	1.1

Table 7.7 Impact of watershed development activities on the village economy (Rs. in Lakhs)

Crop	TS	CS	PS
Maize	75.6	26.1 (34.5)	49.5 (65.5)
Sorghum	37.8	18.5 (48.9)	19.3 (51.1)
Cotton	461.0	192.0 (41.6)	269.0 (58.4)
Rice	97.0	38.0 (39.2)	59.0 (60.8)
Pigeon pea	95.0	35.0 (36.8)	60.0 (63.2)
Vegetables	80.0	35.0 (43.8)	45.0 (56.3)
Chickpea	74.0	25.0 (33.8)	49.0 (66.2)

Table 7.8 Results of economic analysis employing economic surplus method

Particulars	Values
BCR	2.26
NPV (Rs in lakhs)	141.90
IRR (%)	31

interventions. The availability of water has influenced farmers to cultivate cash crops such as cotton and vegetables, and rice is being cultivated as staple crop.

The change in total surplus due to watershed interventions was decomposed into change in consumer surplus and change in producer surplus. Table 7.7 describes that the producer surplus was higher than consumer surplus in all instances. In absolute terms, the highest producer surplus was observed in case of cotton (269 lakh INR) followed by pigeon pea (60 lakh INR) and rice (59 lakh INR).

The overall impact of watershed activities was assessed in terms of NPV, BCR and IRR. The NPV, BCR and IRR were worked out using the economic surplus approach assuming 10% discount rate. The BCR is worked out to be more than 2, implying that the returns to public investment such as watershed development activities were feasible and economically remunerative. Similarly, the IRR worked out to be 31%, which is higher than the long-term loan interest rate by commercial banks, indicating the worthiness of the government investment on watershed program. The NPV worked out to be Rs. 141 lakh INR for the entire watershed. The total treated area in the watershed was around 465 ha, and the NPV per ha worked out to be Rs. 30,000 INR implied that the benefits from watershed development were higher than the cost of investment of the watershed development programs (Table 7.8).

7.5 Conclusions

Integrated watershed management program at Kothapally contributed significantly in terms of improving rural livelihood. The study revealed that the watershed development has the potential for poverty reduction by generating impressive returns on investment even during drought year. The rainwater harvesting through low-cost water harvesting structures ensured water availability which resulted in improved crop productivity. The low-cost water harvesting structures ensured groundwater recharge and, hence, expansion of irrigated area. The new generation watershed intervention emphasizes achieving the food and income security of farmers while maintaining the integrity of the eco-hydrology and other natural systems in the watershed. Increased water availability resulted in crop diversification using high-value crops like vegetables and Bt cotton. Cotton crop is the major commercial crop grown in this region fetching higher market price and profits as well. The harvesting of cotton along with vegetables gave much higher returns to land and labour compared to other crops. This implies that efforts that can increase physical yields of food and commercial crops would result in tremendous financial earnings in the watershed.

The study applied economic surplus approach to capture the impact of watershed development which can assess the impact of watershed development activities in a holistic manner and assesses the distributional effects of the watershed interventions. The overall impact of watershed activities assessed in terms of NPV, BCR and IRR revealed that the BCR is more than 2, implying that the returns on investment were feasible. Similarly, the IRR and NPV were worked out to be 31% and Rs. 141 lakh INR indicating the worthiness of the government investment on watershed program. The study results have demonstrated that the watershed development program has been the potential engines for improving the livelihood of small farm holders along with addressing environmental security.

References

Anantha, K. H., & Wani, S. P. (2016). Evaluation of cropping activities in the Adarsha watershed project, southern India. *Food Security, 8*(5), 885–897. ISSN 1876–4517.

Chandrakanth, M. G., Alemu, B., & Bhat, M. G. (2004). Combating negative externalities of drought–groundwater recharge through watershed development Programme. *Economic and Political Weekly, 13*(March), 1164–1170.

Deshpande, R. S., & Narayanamoorthy, A. (1999). An appraisal of watershed development programme across regions in India. *Artha Vijnana, 41*(4), 415–515.

Garg, K. K., & Wani, S. P. (2012). Opportunities to build groundwater resilience in the semi-arid tropics. *Ground Water*. 1–13. ISSN 1745-6584.

Garg, K. K., Karlberg, L., Barron, J., Wani, S. P., & Rockstrom, J. (2011). Assessing impacts of agricultural water intervention in the Kothapally watershed, Southern India. *Hydrological Process*. https://doi.org/10.1002/hyp.8138.

Government of India. (2001a). *Guidelines for watershed development*. (Revised 2001). New Delhi: Department of Land Resources, Ministry of Rural Development, Government of India.

Government of India. (2001b). *Report of the working group on watershed development, rainfed farming and natural resource management for the tenth five year plan*. New Delhi: Planning Commission, Government of India.

Government of India. (2007). *Report of the working group on natural resources management, eleventh five year plan (2007–2012), volume I: Synthesis*. New Delhi: Planning Commission, Government of India.

Government of India. (2008). *Common guidelines for watershed development projects.* New Delhi: Department of Land Resources, Ministry of Rural Development, Government of India.

IEG World Bank. (2008). Independent Evaluation Group. 2008. *Annual Review of Development Effectiveness 2008: Shared Global Challenges.* Washington, DC: World Bank © World Bank.

Joshi, P. K., Jha, A. K., Wani, S. P., Joshi, L., & Shiyani, R. L. (2005). Meta-analysis to assess impact of watershed program and people's participation. In *Research report 8, comprehensive assessment of watershed management in agriculture* (p. 21). International Crops Research Institute for the Semi-Arid Tropics and Asian Development Bank.

Joshi, P. K., Jha, A. K., Wani, S. P., Sreedevi, T. K., & Shaheen, F. A. (2008). *Impact of watershed program and conditions for success: A meta-analysis approach* (Global theme on agroecosystems report no. 46, p. 24). Patancheru: International Crops Research Institute for the Semi-Arid Tropics.

Kerr, J., Pangare, G., Pangare, V. L., & George, P. J. (2000). *An evaluation of dryland watershed development in India* (EPTD discussion paper 68). Washington, DC: International Food Policy Research Institute.

Libardo, R. R., García, J. A., Seré, C., Jarvis, L. S., Sanint, L. R., & Pachico, D. (1999). *Manual on economic surplus analysis model (MODEXC).* Colombia: International Centre for Tropical Agriculture.

Maredia, Mywish, Byerlee, Derek, Anderson, Jock (2000) Ex-post evaluation of economic impacts of agricultural research programs: A tour of good practice. Paper presented at the Workshop on The Future of Impact Assessment in CGIAR: Needs, Constraints, and Options, Standing Panel on Impact Assessment (SPIA) of the Technical Advisory Committee, Rome, 3–5, May.

Moore, M. R., Gollehon, N. R., & Hellerstein, D. M. (2000). Estimating producer's surplus with the censored regression model: An application to producers affected by Columbia River basin Salmon recovery. *Journal of Agricultural and Resource Economics, 25*(2), 325–346.

Palanisami, K., & Kumar, Dr. (2009). Impacts of Watershed Development Programmes: Experiences and Evidences from Tamil Nadu. *Agricultural Economics Research Review, 22,* 387–396.

Palanisami, K., Suresh Kumar, D., Wani, S. P., & Mark, G. (2009). Evaluation of watershed development Programmes in India using economic surplus method. *Agricultural Economics Research Review, 22,* 197–207.

Pachico, D., Lynam, J. K., & Jones, P. G. (1987). The distribution of benefits from technical change among classes of consumers and producers: An ex-ante analysis of beans in Brazil. *Research Policy, 16,* 279–285.

Pathak, P., Wani, S. P., Sudi, R., Chourasia, A. K., Singh, S. N., & Kesava Rao, A. V. R. (2007). *Rural prosperity through integrated watershed Management: A case study of Gokulpura-Goverdhanpur in Eastern Rajasthan* (Global theme on agroecosystems report no. 36). Patancheru: International Crops Research Institute for the Semi-Arid Tropics.

Sreedevi, T. K., Shiferaw, B., & Wani, S. P. (2004). *Adarsha watershed Kothapally: Understanding the drivers of higher impact.* Patancheru: International Crops Research Institute for the Semi-Arid Tropics.

Sreedevi, T. K., Wani, S. P., Sudi, R., Patel, M. S., Jayesh, T., Singh, S. N., & Shah, T. (2006). *On-site and off-site impact of watershed development: A case study of Rajasamadhiyala, Gujarat, India* (Global theme on agroecosystems report no. 20). Patancheru: International Crops Research Institute for the Semi-Arid Tropics.

Swinton, S. M. (2002). Integrating sustainability indicators into the economic surplus approach for NRM impact assessment. In B. Shiferaw & H. A. Freeman (Eds.), *Methods for assessing the impacts of natural resources management research. A summary of the proceedings of the ICRISAT-NCAP/ICAR international workshop.* Hyderabad: ICRISAT, Patancheru, 6–7 December.

Wander, A. E., Magalhaes, M. C., Vedovoto, G. L., & Martins, E. C. (2004). Using the economic surplus method to assess economic impacts of new technologies — Case studies of EMBRAPA,

Rural Poverty Reduction through Research for Development, Conference on International Agricultural Research for Development, Deutscher Tropentag, 5–7,October, Berlin.

Wani, S. P., & Ramakrishna, Y. S. (2005). Sustainable Management of Rainwater through integrated watershed approach for improved rural livelihood. In B. R. Sharma, J. S. Samra, C. Scott, & S. P. Wani (Eds.), *Watershed management challenges: Improved productivity, resources and livelihoods*. Sri Lanka: International Water Management Institute.

Wani, S. P., & Johan, R. (2011). *Integrated watershed management in rainfed agriculture*. The Netherlands: CRC Press/Balkema. ISBN: 978-0-415-88277-4.

Wani, S. P., Singh, H. P., Sreedevi, T. K., Pathak, P., Rego, T. J., Shiferaw, B., & Iyer, S. R. (2003). *Farmer-participatory integrated watershed management: Adarsha watershed, Kothapally, India: An innovative and Upscalable approach*. Patancheru: International Crop Research Institute for the Semi-Arid Tropics.

Wani, S. P., Joshi, P. K., Raju, K. V., Sreedevi, T. K., Wilson, J. M., Amita, S., Diwakar, P. G., Palanisami, K., Marimuthu, S., Jha, A. K., Ramakrishna, Y. S., Meenakshi Sundaram, S. S., & D'Souza, M. (2008). *Community watershed as a growth engine for development of dryland areas. A comprehensive assessment of watershed programs in India* (Global theme on agro-ecosystems report no. 47, p. 36). Patancheru: International Crops Research Institute for the Semi-Arid Tropics.

Wani, S. P., Sahrawat, K. L., Sarvesh, K. V., Mudbi, B., & Krishnappa, K. (Eds.). (2011a). Soil fertility atlas for Karnataka, India (p. 312). Patancheru: International Crops Research Institute for the Semi-Arid Tropics (ICRISAT). ISBN 978-92-9066-543-4. Order code: BOE 055.

Wani, S. P., Anantha, K. H., Sreedevi, T. K., Sudi, R., Singh, S. N., & D'souza, M. (2011b). Assessing the environmental benefits of watershed development: Evidence from the Indian semi-arid tropics. *Journal of Sustainable Watershed Science and Management, 1*(1), 10–20.

Wani, S. P, Ramakrishna, Y. S., Sreedevi, T. K., Long, T. D., Thawilkal, W., Shiferaw, B., Pathak, P., & Kesava Rao, A. V. R. (2006). *Issues, concepts, approaches and practices in the integrated watershed management: Experience and lessons from Asia in Integrated Management of Watershed for Agricultural Diversi1cation and Sustainable Livelihoods in Eastern and Central Africa: Lessons and experiences from semi-arid South Asia* (pp. 17–36). Proceedings of the International Workshop held 6–7 December 2004 at Nairobi, Kenya.

Wani, S., & Shiferaw, B. (2006). *Baseline characterization of benchmark watersheds in India*. Journal of SAT Agricultural Research: Thailand and Vietnam.

Wani, S. P., Sreedevi, T. K., Rockström, J., & Ramakrishna, Y. S. (2009). Rain-fed agriculture past trend and future prospects. In rain-fed agriculture: Unlocking the potential. In S. P. Wani, J. Rockström, & T. Oweis (Eds.), *Comprehensive assessment of water Management in Agriculture Series* (pp. 1–35). Wallingford, UK: CAB International.

Chapter 8
Digital Technologies for Assessing Land Use, Crop Mapping and Irrigation in Community Watersheds

V. R. Hegde and K. V. Raju

Abstract Land-use analysis, cropping pattern and sources of irrigation are important for planning and improving the rural economic aspects. Geospatial method was adopted for deriving and analysing the land use at micro-level in Kothapally Village covering parts of Adarsha watershed. The village is characterised by a large number of marginal holdings, and the average holding of 0.96 ha. is lesser than that in Telangana state. Kothapally is characterised by agriculture, and the land put to non-agricultural use is 9.31%. Migration to urban areas, selling road side land for commercial purposes and alienated lands not being cultivated by the poor have been found to be the main reasons for fallow lands. Cropping systems comprise of cereals, pulses, fruits, commercial crops, vegetables, flowers and plantations. Over the years' extent of cereals have remained almost same, with sorghum cultivation almost stopped. Pigeon pea appears to be gaining prominence and area under cotton has doubled. Sugarcane has disappeared.

Rainfall is the only source of irrigation in Kothapally. Groundwater is being used for irrigating 109.49 ha. Groundwater withdrawal is on the rise and shallow open wells are becoming unusable. Different types of rainwater harvesting have been adopted. Though interventions have yielded better results, more demand for water has caused more exploitation of groundwater. The village has all the required basic infrastructural facilities, and the wastewater treatment facilities built as part of watershed management interventions are appreciated by the community. Almost all the houses have sanitary facilities and are connected with underground drainage.

Keywords Land use · Marginal holdings · Cropping pattern · Pigeon pea · Cotton · Sugarcane

V. R. Hegde (✉)
Pixel Softek Pvt. Ltd, Bengaluru, Karnataka, India
e-mail: vrhegde@pixelsoftek.com

K. V. Raju
Former Theme Leader, Policy and Impact, Research Program-Asia, International Crops Research Institute for the Semi-Arid Tropics (ICRISAT), Hyderabad, Telangana, India

© Springer Nature Switzerland AG 2020   143
S. P. Wani, K. V. Raju (eds.), *Community and Climate Resilience in the Semi-Arid Tropics*, https://doi.org/10.1007/978-3-030-29918-7_8

8.1 Introduction

Land, besides being a resource, is also a resource base. Land refers not only to soil but also landforms, climate and hydrology, plant and animal population and the physical results of human activity (Sombroek 1992, FAO). Land is considered to be the most important component of earth system and being used with an increasing intensity to meet the needs of a growing population. These needs relate to increasing demands for food and space and better material expectations also. It is estimated that the human footprint has affected 83% of the global terrestrial land surface and has degraded about 60% of the ecosystems services in the past 50 years alone. Land-use and land-cover (LULC) change has been the most visible indicator of the human footprint and the most important driver of loss of biodiversity and other forms of land degradation (SD21 2012).

Land cover is fundamental and is the observed biophysical cover, a geographically explicit feature on the earth's surface, while the land use is characterised by the arrangements, activities and inputs people undertake in a certain land-cover type to produce, change or maintain it. Definition of land use in this way establishes a direct link between land cover and the actions of people in their environment. Land use is the surface utilisation of all developed and vacant land on a specific point, at any given time and space (Mandal 1990). While it defines human activities, the emphasis is on the purpose for which the land is used, and particular reference is made to "the management of land to meet human needs" (FAO 1976). The land-use pattern of a region is an outcome of natural and socioeconomic factors which decide the utilisation of land by human beings over time and space. All agricultural, animal and forestry productions depend on the productivity of the land. The entire ecosystem of the land, which comprises of soil, water and plant, meets the community demand for food, feed, energy and other needs of livelihood.

In rural landscape, life and productive activities depend on natural resources and environmental conditions. Between the rural settlements and the surrounding landscape, a close relationship exists in terms of morphology and economic life. In rural India, villages are defined with a clear administrative boundary and land holdings, forming a specific ecosystem with finite resources that are subject to changes and variations and thereby influence the land use. The terrain, quality of soil and its management control the agriculture. As communities directly depend on land and water resources, land use has specific consequence on the services of the ecosystem. The spatial dimensions of land use help in understanding resources use and economic and social conditions.

8.1.1 Land-Use Assessment

The land-use assessment concept has been promoted mainly to encourage the proper use of land and also to conserve natural resources. The conservation-based ideology has taken a paradigm shift towards development-oriented concepts after The Earth

Summit in Rio de Janeiro (1992) and the publication of Agenda 21 on sustainable development concepts in rural areas. The new perceptions are on planning and management of land resources keeping in view the economic, environmental, social and institutional aspects (De Wit and Willy 2009). Sustainable agriculture is being advocated as an approach to securing the necessary resources for safeguarding global food production, biodiversity reserves, recreation needs, water quality and well-developed rural areas. It can also be an effective means of poverty reduction and of achieving the Millennium Development Goals (MDGs) and now Sustainable Development Goals (SDGs), as well as means of mitigating climate change. In other words, they are truly transdisciplinary and represent a new holistic outlook on ecosystem health and sustainable agriculture (Karlsson and Rydén 2012).

Agriculture and related land-use change continuously in response to the market forces, and general awareness has arisen about the effects of these changes (Bernetti et al. 2006). The development of the agricultural sector has included structural transformation of farms, as well as land-use changes, to meet market requirements in terms of economic efficiency. These leading forces have trapped agricultural land between the phenomena of specialisation/intensification and abandonment of higher cost and less competitive production areas. These two distinct phenomena are taking place in the context of the complex interaction between biophysical and socio-economic factors operating at various levels and driving land-use pattern modifications with implications for the multifunctionality of agriculture (Bernetti and Marinelli 2010).

Urbanisation has also been one of the main causes that has been exerting substantial pressure on rural ecosystems. Most of the rural landscapes in the semiarid tropical regions and particularly the Peninsular India have been facing challenges of climate (mostly the vagaries of monsoons) as most of the river systems are dependent on rainfall. In addition to the climatic variations, urbanisation also has been one of the reasons for changing land use in this part of the country. A study on the influence of development of Bangalore has brought out effects of compounded policies, urbanisation compulsion and strategic actions of economic growth and response of the system in the surrounding rural areas. Sen (2016) in relation to Hosakote, identifying the rural-urban fringe with characteristic land-use associations, opined that a conflict of two lifestyles and direct impact of urban expansion on agricultural lands with clear indications of urban elements exist in peri-urban area.

Land-use assessment in rural areas is becoming more important and a necessity so that development of local ecological interventions is used for increasing the productivity in a sustainable manner. Micro-planning is of critical importance at the village level mainly to ascertain the needs and aspirations of people in rural areas. The Constitution of India has mandated all rural communities have powers to govern themselves through elected panchayats and plan for economic development and social justice in their jurisdictions (GOI). Studying farming systems in a village allows one to address proximate causes and to interpret them in reference to underlying causes. Underlying causes can be interpreted as "individual and social responses to changing economic and technological conditions, which are mediated by institutional factors" and are "formed by a complex of social, political, economic, demographic, technological, cultural, and biophysical variables" (Lambin et al. 2003).

8.1.2 Land-Use Classification

Land-use classification is of great significance as it is the most fundamental step of land-use analysis. Land-use classification is the systematic arrangement of various classes of land on the basis of certain similar characteristics, mainly to identify and understand their fundamental utilities, intelligently and effectively in satisfying the needs of the human society. The purpose of land-use classification is to maximise the productivity and conserve the land for posterity (Mandal 1990). According to Barnes (1936), the objectives of land-use classification can be categorised into the following:

- More enlightened and economically sound land settlement policies, both public and private
- Guidance in public land purchase and development
- Planning the organisation and distribution of local government services
- Guidance in the distribution of public aids
- Guidance in determining sound real estate lending and borrowing policies
- Land assessment for taxation purposes
- Developing administration programmes for land conservation and management
- Developing sound farm management policies and organizing the most effective
- Deciding the type and size of operative units

The development of land classification systems has a long history in various countries of the
world. There are discussions on land-use classifications in *Sangam literatures*. In that, general land use was classified into five categories: (1) *Kurunji* (hilly region), (2) *Mullai* (forest region), (3) *Marutham* (arable land), (4) *Neithal* (coastal region) and (5) *Palai* (desert region). It was more oriented towards classifying landscapes and land-cover types. Soil classification systems were the first to be produced by both national (e.g. the US Soil Conservation Service, Canada's Soils Directorate) and international (the FAO) organisations to serve the needs of producing soil maps and provide a basis for determining land capability and suitability for growing various types of crops (Marsh 1991). Land-use classification is an abstract representation of the situation in the field using well-defined diagnostic criteria, and it is the arrangement of objects into groups or sets on the basis of their relationships (Sokal 1974). Classification requires the definition of class boundaries, which should be clear, precise, possibly quantitative and based upon objective criteria and therefore be:

- Scale independent, meaning that the classes should be applicable at any scale or level of detail
- Source independent, implying that it is independent of the means used to collect information, whether it be through satellite imagery, aerial photography, field survey or a combination of sources

Land-use classification system developed by the US Geological Survey (USGS) has multiple levels of classification. This level of classification is commonly used for regional and other large-scale applications. Within each of the level I classes are a number of more detailed (level II) land-use and land-cover classes. Within each of the level II classes, even more detailed classes (levels III and IV) can be defined and mapped. The classes within each level are mutually exclusive and exhaustive. That is, each location within the mapped area can be classified into one and only one class within each level (LaGro 2005). According to Stamp (1951), one of the pioneer of land-use studies and who identified six classes of land use for Britain, an ideal land-use classification system should be exhaustive so that it does not omit any phenomenon. It should also be mutually exclusive so that the categories identified do not overlap with each other.

Despite the need for a standard classification system, none of the current classifications have been internationally accepted (Fosberg 1961; Eiten 1968; UNESCO 1973; Mueller-Dombois and Ellenberg 1974; Kuechler and Zonneveld 1988; CEC 1993). This is because often the land-cover classes have been developed for a specific purpose or scale and are therefore not suitable for other initiatives. Further, factors used in the classification system often result in a mixture of potential and actual land cover. Most of the existing classifications are either vegetation classifications, broad land-cover classifications or systems related to the description of a specific feature (Di Gregorio and Jansen 2000). A classification system should be based on diagnostic criteria and suitable for mapping, not having conflict between classes, and should be with definite boundary between classes, and definition should be consistent. Further, the classification shall allow monitoring of changes from one category to another.

8.1.3 Land-Use Classification and Mapping in India

The evolution of land-use statistics in India dates back to 1866 when the British administration took interest in the compilation of land data to enhance its revenue collection (Suresh Kumar 2014). While compiling the historical land use data related to croplands, forests, grasslands/shrublands and built-up areas at state level for the period 1880–1980 from the archives, Richards and Flint (1994) observe that though there are certain limitations of the inventory of land use datasets, quality of data seems to be improving since the governments were interested in enumerating productive assets, especially land used for crops and forests. The statistical system of the erstwhile British era identified five broad indicators like (1) forest, (2) area not available for cultivation, (3) other uncultivated land excluding current fallows, (4) fallow land and (5) net sown area. After independence, a Technical Committee on Coordination of Agricultural Statistics constituted by the Ministry of Agriculture identified major gaps in the existing data collection on agriculture and suggested to add four more categories to make it a ninefold classification of the total land available. Since then, ninefold classification is followed in India.

Out of a geographical area of 329 million hectares (reporting area), statistics are available only from 305 million hectares, which makes some areas to the extent of 7% still not covered or classifiable under the ninefold classification (MOSPI). The reporting area is classified into the following nine categories:

- **Forests**: This includes all lands classed as forest under any legal enactment dealing with forests or administered as forests, whether state-owned or private and whether wooded or maintained as potential forest land. The area of crops raised in the forest and grazing lands or areas open for grazing within the forests should remain included under the forest area.
- **Area under non-agricultural uses**: This includes all lands occupied by buildings, roads and railways or under water, e.g. rivers and canals, and other lands put to uses other than agriculture.
- **Barren and uncultivable land**: This includes all barren and uncultivable land like mountains, deserts, etc. Land which cannot be brought under cultivation except at an exorbitant cost should be classed as uncultivable whether such land is in isolated blocks or within cultivated holdings.
- **Permanent pastures and other grazing lands**: This includes all grazing lands whether they are permanent pastures and meadows or not. Village common grazing land is included under this head.
- **Land under miscellaneous tree crops, etc.**: This includes all cultivable land which is not included in "Net sown area" but is put to some agricultural uses. Lands under *Casuarina* trees, thatching grasses, bamboo bushes and other groves for fuel, etc. which are not included under "Orchards" should be classed under this category.
- **Cultivable wasteland**: This includes lands available for cultivation, whether not taken up for cultivation or taken up for cultivation once but not cultivated during the current year and the last 5 years or more in succession for one reason or other. Such lands may be either fallow or covered with shrubs and jungles, which are not put to any use. They may be assessed or unassessed and may lie in isolated blocks or within cultivated holdings. Land once cultivated but not cultivated for 5 years in succession should also be included in this category at the end of the 5 years.
- **Fallow lands other than current fallows**: This includes all lands, which were taken up for cultivation but are temporarily out of cultivation for a period of not less than 1 year and not more than 5 years.
- **Current fallows**: This represents cropped areas, which are kept fallow during the current year. For example, if any seeding area is not cropped against the same year, it may be treated as current fallow.
- **Net sown area**: This represents the total area sown with crops and orchards. Area sown more than once in the same year is counted only once.

Information on land use in the form of maps and statistical data is very important for spatial planning, management and utilisation of land. Realizing the need for an up to date nationwide land use/land cover maps by several departments in the country, as a prelude, a land use/land cover classification system (with 24 categories up to Level-II, suitable for mapping on 1: 250,000 scale) was developed by NRSA, DOS, taking into consideration the existing land use classification adopted by NATMO, CAZRI, Ministry of Agriculture, Revenue Department, AIS and LUS etc. and the details obtainable from satellite imagery. The classification system provided the conceptual framework for discussions with nearly 40 user departments/institutions in the country and finalises acceptable 22-fold classification system which was adopted for nationwide land-use/land-cover analysis (Roy and Giriraj 2008).

8.2 Studies on Land-Use Characterisation

There is no one ideal classification of land use and land cover, and different schema are being used to suit the needs of the user and constantly changing patterns keeping with demands for natural resources (Anderson et al. 1976). Most important aspect is the size of the minimum area which can be depicted as being in any particular land-use category depends partially on the scale. In the recent years due to advancements in the spatial resolution of the sensor on board, numerous studies are being conducted. Many of them have used multi-date images for characterising the changing patterns of land use (Halder 2013; Chattopadhyay et al. 1999; Premakumar and Vinothkanna 2015; Kannadasan et al. 2017; Tian et al. 2014). Neural networks approach has been adopted for land-use characterisation using satellite data (Castelluccio et al. 2015).

Since cropland constitutes important class in any land-use study, focus also has been on mapping of agricultural lands. Remote-sensing techniques have become popular in crop area mapping over the past few decades, and as the technology and methodologies have matured, major research and development thrust has been on agricultural crop identification and area estimation (Dadhwal et al. 2002). Geospatial technologies (remote sensing and geographic information system) have been used to assess the agricultural potential of the Nebo Plateau, a rural area in the Limpopo Province of South Africa (Brilliant Petja et al. 2014). The approach entails assessing the suitability in terms of land/soil and climate, which are determinant factors for agricultural development.

8.2.1 Land-Use Information at Micro-Level

In the rural context, the landscape with all its assets, including natural resources, is the key resource to take care of in planning and to manage properly. Without a well-managed and productive landscape, the rural society will be bereaved of its key resource and indeed its condition for long-term survival (Kristina Nilsson and Rydén 2012). Governments, policies as well as global and domestic markets set the conditions, under which micro-agents, i.e. households, firms and farms, eventually take and implement decisions on land use. By placing an emphasis on the micro-level studies, one can provide a more detailed assessment of household-level factors for land-use change and the heterogeneity in the relationship between land-use change and growth-associated micro-level (Hettig et al. 2016). Land-use studies, therefore, in rural ecosystem need to focus more on the agriculture classes and should be able to correlate with farm types and land holdings. This would facilitate the local governments to devise local-level plans for sustainable development conforming to the policies at the state level. Miller et al. (2012) while determining the land-use and farm-type mixes of land capability for agriculture mention that land-use assessment at holding level is of paramount importance to support the development of options for area-based farm payments. The changing spatial patterns of land use including crop diversity provide information on the drivers and implications of changes in rural area (Tiwari et al. 2010). Crop mapping at plot and/or holding level becomes most useful, and a combination of remote sensing and other geospatial tools was found to be more reliable (Naik et al. 2013).

Geospatial technologies have been recognised as the tools that strengthen the quality of geographical data in terms of both reliability and in-built capacity of traceability. The use of satellite images has also been explored for crop mapping, and it is found that regions with fragmented holding and mixed cropping system and remote-sensing techniques have certain limitations. Cloud coverage is one of the factors that hinders the use of remote sensing, and further, in a mixed cropping system, it becomes difficult to delineate different crops. The use of handheld GPS also might pose certain constraints due to signals from orbiting GPS satellites.

ICRISAT has adopted an integrated method for crop enumeration in different agro-climatic regions of India. The approach named "geo-stamping" was deployed for generation of Crop-Area Statistics of Khatijapura (Karnataka), Daliparru (Andhra Pradesh) and Jarasingha (Odisha) villages. The strategic interventions have been the use of high-resolution satellite data for extracting crop plots, GIS for spatial data integration and visualisation, GPS for recording relevant details of crop and sources of irrigation on the spot, etc. Based on the mapping of cropping pattern and other land-use details at plot level, land use for the entire village was deduced.

8.3 Kothapally Village

Kothapally Village (Adarsha watershed forming a part) is located in Kothapally Village (longitude 78°5′ to 78°8′ E and latitude 17°20′ to 17°24' N) of Shankarpally Mandal in Ranga Reddy district, Telangana, India, nearly 40 km from ICRISAT, Patancheru (Fig. 8.1).

8.3.1 Methodology

Micro-level land-use studies need to have detailed assessment of household-level factors for land use, focussing on agriculture classes and farm types and holdings. Existing methods either remote sensing or sampling do not provide the details

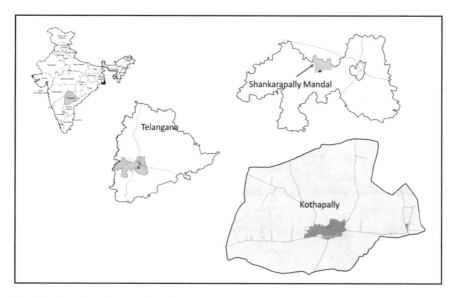

Fig. 8.1 Location of Kothapally Village

required for local-level planning. Kothapally is a small village panchayat, and characterisation of land use at holding level is necessary for the following reasons:

- Entire village is under rain-fed agriculture.
- Characterised by marginal to small holdings.
- ICRISAT has been implementing various interventions for building up natural resources for the past two decades in parts of the village.

In view of the above considerations, it was decided to adopt geo-stamping approach that was tested and established by ICRISAT for similar micro-level studies in different parts of the country.

The following are the pre-field activities:

- The village Revenue Survey Map of Kothapally was vectorised, and all the survey number boundaries (Fig. 8.2) were organised in GIS environment and proper codification was done.

- The high-resolution satellite image (Google Earth) was deployed in GIS environment, and crop plots as discernible from the image were extracted, and unique codes were generated for each of the crop plot in association with the respective full survey number. Similarly, the Kothapally settlement map was prepared wherein each household was provided with a unique number (Figs. 8.3 and 8.4).

- A controlled mosaic of crop plots for the entire village was developed in GIS environment, and geo-fenced map tile for each village was generated (Fig. 8.5).

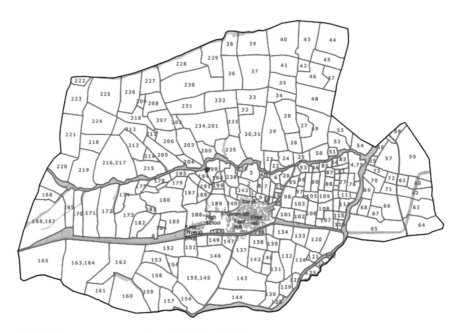

Fig. 8.2 Revenue survey numbers in Kothapally Village

Fig. 8.3 Google Earth image used for mapping crop plots and households in Kothapally

- A comprehensive Android application was developed (Figs. 8.6 and 8.7), and the data collection format decided by the project team of ICRISAT was implemented in addition to standard field mapping parameters such as crop and land use and house-hold data.

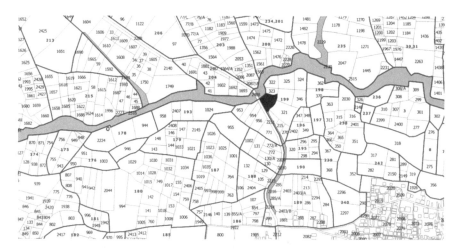

Fig. 8.4 Crop plots identified within each survey number in Kothapally

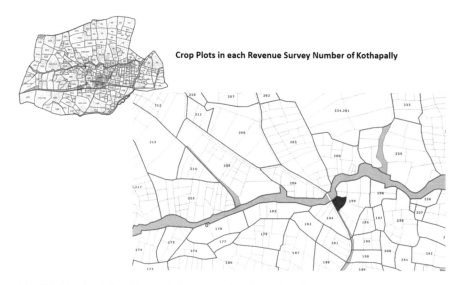

Fig. 8.5 Mosaic of Crop Plots with Survey numbers in Kothapally

- The geo-fenced map tile was loaded to GPS-based smartphones, and the application was tested for ease of data authentication and recording of geo-coordinates of each crop plot and other information on the spot along with photographs. The application has the following functionalities:

  - Automatic loading of the village map tile on the screen in the field and displaying of location of the enumerator
  - Capturing of the location (coordinates) of the plot (registering the plot)

Fig. 8.6 Screenshots of app used for crop inventory

Fig. 8.7 Screenshots of app used for household survey

- Recording details of the crop in the plot
- Capturing of photograph of the crop
- Capturing the coordinates of features like borewells and any other structures in the land
- Capturing other information (as specified by project team of ICRISAT)
- Downloading the data and synchronising with the GIS database residing in the backend system

8.3.2 Field Session and Supporting Work

Field inventory of land use at plot level in Kothapally was conducted during November 2017. The season corresponds to the later part of *kharif* (rainy season) of 2017–2018. Totally ten members forming five teams were deployed. To begin with, crop inventory was carried out. It was followed by household survey.

- Issues during the survey:
  - During the field inventory, though most of the farmers were available on the spot for confirmation of their land, few had gone out of village. In such cases, the ownership was confirmed by the neighbouring farmer, and that was correlated with the records from the gram panchayat.
  - Similarly, during the household inventory, few residents were not available in the first instance. The team had to revisit such houses for getting correct information.
- The data received from each GPS unit from the field on daily basis was maintained in separate folders.
- The attribute information of each crop plot was integrated, and a special team for quality checking was deployed for the analysis of recording errors, if any, like name of crop.
- All the data from GPS devices were systematically ported to the basic GIS database organised on the open-source GIS platform, i.e. QGIS. Different layers for specific crops and other base details were created, and the area statistics were generated (Fig. 8.8).

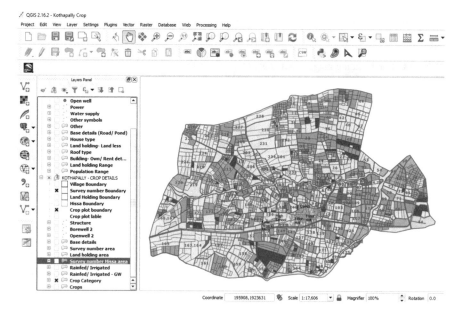

Fig. 8.8 Spatial and attribute data organised in QGIS

8.4 Land Holding

The field inventory conducted at individual crop plot level was compiled systematically for understanding the agricultural holding pattern in Kothapally. The standard land holding classification (PIB 2015) has been adopted. The land holding pattern in Kothapally is shown in Table 8.1 and Fig. 8.9.

The village has 524 holdings of varying sizes. The average holding is 0.96 ha. Kothapally is characterised by large number of marginal holdings, i.e. 399 holdings with less than 1 ha. The field survey has also revealed many combined holdings with more than one *Khata* without any field demarcation (Fig. 8.10). The data maintained at GP does not tally with the current scenario, mainly because the data has not been updated. In the revenue records, the data is updated (though at places field demarcation is not done). Further, during the field inventory, it was learnt that substantial extent has been alienated to the poor, but the *khata* still remains in the name of the government in the records. GP has not maintained the records as per the land holding classification also, and therefore, the books do not show few categories of holdings.

Land holding is a significant input for agriculture. The average land holding in Telangana in 2010–2011 was 1.12 ha (2.8 acres) against the all India average of 1.16 ha, as per the Agriculture Statistics of Telangana 2015–2016. The average size of holdings has shown a steady declining trend over various agriculture censuses since 1970–1971. Increase in population has put pressure on land, leading to fragmentation of holdings. The share of small and marginal land holdings constitutes about 86% of total land in the state (Table 8.2), while their share in total area was only 55% in 2010–2011. About 14% of total land holdings in the state were medium, ranging between 2 and 10 ha, whereas their share in total area was 40.5% (TNE 2018).

Table 8.1 Land holding range in Kothapally

Sl. no.	Land holding types	As per field inventory			As per GPS[a] records
		No. of holdings	%	Area in Ha	No. of holdings
1	Marginal holdings (<1 ha)	399	76.1	176.24	253.00
2	Small holdings (1–2 ha)	87	16.6	120.76	151.00
3	Semi-medium holdings (2.1–4 ha)	28	5.3	80.61	
4	Medium holdings (4.1–10 ha)	7	1.3	33.45	
5	Large holdings (>10 ha)	3	0.6	38.95	40.00
6	Non-agricultural land	–		43.34	
7	Govt. land	–		11.20	
8	Total	524		504.55	444.00

[a]Gram Panchayat

Land Holding Range in Ha
Marginal Holdings (< 1 ha)
Small Holdings (1 - 2 ha)
Semi-medium Holdings (2.1 - 4 ha)
Medium Holdings (4.1 - 10 ha)
Large Holdings (> 10 ha)

Fig. 8.9 Land holding pattern in Kothapally

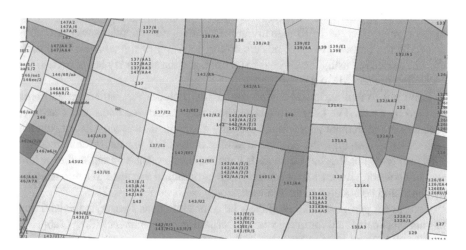

Fig. 8.10 Single piece of land without field demarcation and different *khata*, Kothapally

Table 8.2 Land holdings in Telangana

Category	No. of holdings	Area operated	Holdings % area operated	Area operated %
Marginal holdings	34,41,087	39,16,947.50	61.96	25.28
Small holdings	13,27,362	46,73,380	23.9	30.17
Semi-medium holdings	6,02,925	39,62,837.50	10.86	25.58
Medium holdings	1,66,833	23,16,900	3	14.96
Large holdings	15,775	6,21,997.50	0.28	4.01
Total	55,53,982	1,54,92,060		

Source: TNE, ninth May 2018

8.5   Land-Use Pattern

Land-use classification based on the data gathered during the *kharif* season was derived for Kothapally Village. The standard ninefold land-use classification was considered for compiling the land-use information. The land-use details are presented in Table 8.3 and Fig. 8.11. Except for the substantial extent of land uncultivated for a long period and seasonal fallow land, the net sown area comes to 73.10%, and the land put to non-agricultural use is about 9.31%.

The geographical area of Kothapally has been derived from the village map obtained from the concerned department. Detailed scrutiny of the records revealed small variation between the areas obtained from GIS, which is available in official records. During field inventory, care was taken to ensure that all the lands belonging to the revenue records of Kothapally are registered, and boundary was also confirmed by the farmers (whose lands are on the boundary of the village). Therefore, the area as derived using the GIS has to be used, and the same may be referred to for correcting the records in the Panchayat.

Land put to non-agricultural land essentially includes the settlement and area covered by roads. The GP records have shown a smaller area under this category; probably the settlement area was not properly determined, *or* it must have taken the area under *Gramthan* – the original area of settlement as per the revenue records. This also indicates that the GP records have not been updated based on the current status. Therefore, the current area under non-agricultural use need to be considered the latest and be updated in the records.

Area under permanent fallow (fallow land for more than 2 years) is the land that has not been cultivated for many years. Distribution of permanent fallow (Table 8.4 and Fig. 8.12) and kind of holdings indicate that these are mostly concentrated in the western part of the village. Table 8.3 indicates that all categories of holdings have been left fallow and the semi-medium holdings take slightly bigger share (22.14 ha.). GP records have to be updated as per the current inventory in the field. The permanent fallow land accounts to 11.92% in Kothapally.

Table 8.3 Land use in Kothapally

		As per survey in Kharif 2017		As per GP records	
Code	Land-use classification	Area in ha.	%	Area in ha.	%
1	Forest	0	0	0.00	0.00
2	Land put to non-agricultural uses	46.96	9.31	15.00	3.18
3	Barren and uncultivable land	0	0	0.00	0.00
4	Permanent pastures and other grazing lands	0	0	0.00	0.00
5	Miscellaneous tree crops and other groves, not included in net sown area	0	0	0.00	0.00
6	Culturable waste	0	0	33.20	7.04
7	Fallow land other than current fallow (fallow land more than 2 years)	60.16	11.92	0.00	0.00
8	Current fallow (seasonal fallow)	28.58	5.67	0.00	0.00
9	Net sown area	368.85	73.1	423.40	89.78
Total		504.55	100	471.60	100.00
Water features					
Stream/river		17.69			
Open well/pond/tank		0.37			
Net sown area					
Annual		0			
Kharif		345.57			
Perennial		23.28			
Total		368.85			

Land put to non agricultural uses
Fallow land other than current fallow
Current Fallow (Seasonal fallow)
Net area sown

Fig. 8.11 Land use in Kothapally

Table 8.4 Permanent fallow – holding details

Sl. no.	Landholding range (ha)	No. of holdings	Area (ha)	Farmers living in urban areas
1	Marginal holdings (<1 ha)	22	9.73	9
2	Small holdings (1–2 ha)	14	14.83	2
3	Semi-medium holdings (2.1–4 ha)	8	22.14	4
4	Medium holdings (4.1–10 ha)	2	5.89	0
5	Large holdings (>10 ha)	0	0.00	0
6	Govt. land (distributed to poor)		7.58	
Total		46	60.16	15

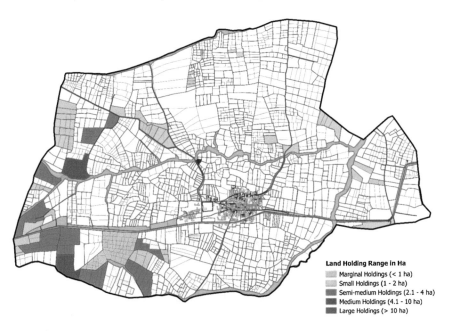

Fig. 8.12 Distribution of permanent fallow lands in Kothapally

During the field inventory, farmers have informed about the probable reasons (Table. 8.5 and Fig. 8.13) for land being left as permanent fallow as follows:

- Farmers who have migrated to urban area
- Road side land being sold to commercial purpose
- The government land distributed to poor not being cultivated
- Underutilised government land
- Other socioeconomic reasons

The government land distributed to poor people for cultivation few years back is 7.58 ha. Since the soil layer is very thin and people have never used it for cultivation, it is lying as fallow land and this needs to be updated in the GP records. Further,

Table 8.5 Reasons for permanent fallow in Kothapally

Sl. no.	Reasons for permanent fallow	No. of holdings	Area (ha)
1	Farmers living in urban areas	14	18.19
2	Land sold to commercial purpose	1	0.76
3	Govt. land distributed to poor	1	7.58
4	Underutilised govt. land	1	1.87
5	Other socioeconomic reasons	30	31.76
Total		47	60.17

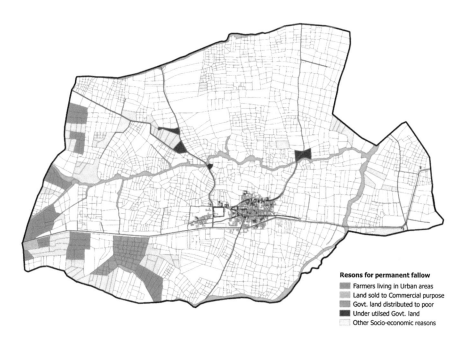

Fig. 8.13 Reasons for permanent fallow and their distribution in Kothapally

few land owners have moved to urban area leaving the land uncultivated. Also, purchase of land for industrial purpose has left land fallow.

Another substantial extent of land not cultivated during *kharif* (2017–2018) is designated as seasonal fallow which accounts to 5.67%. The enquiry during the field inventory indicated that due to certain constraints, farmers have left portion of the land uncultivated during *kharif* of 2017–2018. Distribution of seasonal fallow and the holding types (Table 8.6 and Fig. 8.14) indicate that mostly the marginal farmers have left some plots as fallow. It is also observed that in many cases, a part of the holding is left as seasonal fallow (Fig. 8.15). It is learnt that social causes like

Table 8.6 Current (seasonal) fallow – holding details

Sl. no.	Landholding range (ha)	No. of holdings	Area (ha)	Farmers living in urban areas
1	Marginal holdings (<1 ha)	70	12.51	2
2	Small holdings (1–2 ha)	28	6.79	
3	Semi-medium holdings (2.1–4 ha)	12	5.46	1
4	Medium holdings (4.1–10 ha)	3	1.48	
5	Large holdings (>10 ha)	3	2.35	
6	Total	116	28.58	3

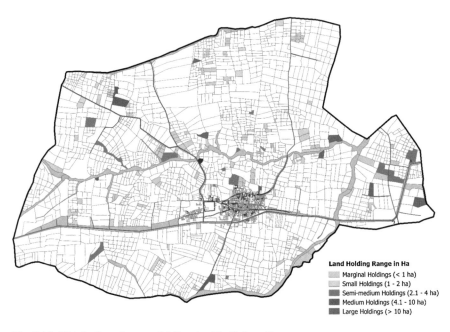

Land Holding Range in Ha
- Marginal Holdings (< 1 ha)
- Small Holdings (1 - 2 ha)
- Semi-medium Holdings (2.1 - 4 ha)
- Medium Holdings (4.1 - 10 ha)
- Large Holdings (> 10 ha)

Fig. 8.14 Distribution of seasonal fallow land in Kothapally

death in the family, economic considerations and in some cases migration to urban area have been the reason.

In case these seasonal fallow lands are cultivated during *rabi* (*post-rainy*), the net sown area would correspondingly increase. During the field inventory, land preparation for *rabi* (post-rainy) crops was observed, and it was also learnt that substantial extent is cultivated during *rabi* (post-rainy) using the groundwater resources. The *rabi* (post-rainy) inventory would have provided data on gross cropped area and also the corresponding (5.67% of the seasonal fallow, if cultivated during *rabi* (post-rainy)) increase of net sown area for the year 2017–2018.

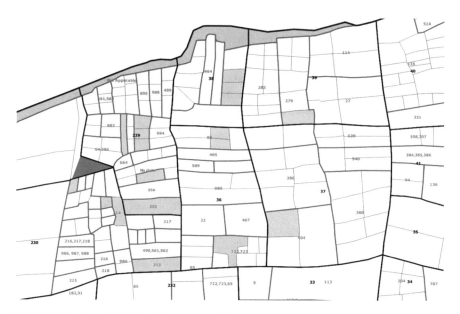

Fig. 8.15 Portion of a holding left as seasonal fallow in Kothapally

The land-use survey at plot level has been useful to understand the complete scenario of cultivation in a village. The GP appears to have maintained the old records and never bothered to update the revenue records. The updated records not only provide the cultivation practice but also provide information about the cropping pattern which will be useful for understanding and devising proper inputs for productivity at micro (village) level. Since rural economy mainly depends on agriculture, land-use information that is being continuously updated would be of great help for planning and developing appropriate market facilities. Further, GP can understand the extent of fallow and reasons thereof and the extent underutilised government land for devising regulatory plans for improving the local economy.

8.6   Crop Mapping

Similar to general land use, the crop statistics are very important, and in a region like Kothapally which is characterised by marginal and small holdings, cultivation is mainly dependent on vagaries of rainfall and the information on cropping pattern, and crop varieties play an important role. Since ICRISAT has been involved in watershed management activities in parts of Kothapally, the cropping pattern assumes significance.

8.6.1   Field Inventory

Before starting the actual field enumeration, a formal discussion with local farmers was held, and farmers were appraised about the work. The team including the scientists from ICRISAT participated in the meeting and has provided useful coordination with the farmers and the functionaries of the Watershed Committee.

8.6.2   The Field Procedure

The following are the specific activities that were carried out in the field:

- The farmers were requested to participate in the enumeration process. They were specifically told to confirm their plots.
- After confirmation of every plot, details like name of the crop grown were recorded in the GPS, and photograph of the crop was taken.
- In case any source of irrigation was seen in the land, the same was captured using GPS.
- Further, details of crop pattern in the field were recorded, i.e. whether it was single crop or crop interspersed with other crop (mixed crop).
- Physical structures like building, sheds, etc. (in case seen in the plot enumerated) were traced.
- Since Internet connectivity was week, the data recorded in the GPS was stored locally in the GPS itself and was downloaded to a computer at the end of the day's session.

8.6.3   Results of Crop Inventory in Kothapally

Results of plot-level crop inventory carried out using GPS-based mobile devices have been compiled and organised in GIS. There is substantial variation in the dimension of the crop plots and crop varieties. In some of the plots, the *kharif* crop was harvested and *rabi* (post-rainy) crops were just started. In such cases, the remnants of earlier crops were recorded as *kharif*, and current status was recorded. The information was also confirmed by the farmer on the spot.

8.6.4   Cropping Pattern

Crop inventory was carried out during November 2017 that corresponds to almost close of the *kharif* season. Each crop plot was geo-stamped and standing crop was recorded (Fig. 8.16). *Kharif* crops noticed in Kothapally are presented in Table 8.7.

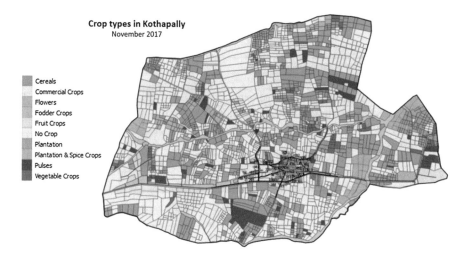

Crop types in Kothapally
November 2017

Cereals
Commercial Crops
Flowers
Fodder Crops
Fruit Crops
No Crop
Plantation
Plantation & Spice Crops
Pulses
Vegetable Crops

Fig. 8.16 Cropping pattern in Kothapally

Table 8.8 shows area occupied by each crop, and Table 8.9 provides the comparison of *kharif* 2017 with *kharif* of earlier years (ICRISAT 2004).

Kothapally is characterised by unique cropping systems comprising of cereals, pulses, fruits, commercial crops, vegetables, flowers and plantations (Figs. 8.17, 8.18, 8.19 and 8.20). Commercial crops occupy larger extent of the crops in Kothapally. Paddy, maize and *jowar* (*sorghum*) are the cereals enumerated and covered an extent of 126.97 ha. Out of these, maize occupies an area of 99.22 ha. Pigeon pea (*tur*) occupies the maximum area (16.57 ha.) amongst the pulses. Cotton was the major commercial crop and occupied an area of 176.24 ha. Different varieties of vegetables were grown in Kothapally, and tomato covered the major area of 5.23 ha. Carnation and rose are the two important flower crops with rose covering an area of 9.1 ha. Fodder crops occupy an area of 4 ha and fruits are grown in an area of 7.35 ha. Mango and orange are almost equally distributed. Plantations occupy an area of 15.93 Ha., and Malabar neem occupies an area of 8.94 ha.

Crop data gathered by ICRISAT earlier (2003–2004) was compared with data generated during November 2017 mainly to understand the changing patter of crops (Table 8.4 and Figs. 8.21, 8.22, 8.23 and 8.24). It appears from the data that there is a gradual shift towards commercial, fruit and plantation over the years.

The data indicate that extent of cereals has reduced and sorghum (*jowar*) cultivation has almost been stopped. Area under maize has increased and paddy has slightly reduced. Amongst the pulses, *tur* (pigeon pea) appears to be gaining prominence since its area has almost doubled compared to earlier records. Under the commercial crops, sugarcane has disappeared and area under cotton has doubled. Cultivation of vegetables has also reduced. During the year, there was no cultivation of any variety of tubers and oil seeds. Significantly, plantations have received attention and Malabar neem and palm cover considerable area in Kothapally.

Table 8.7 Abstract of crops in Kothapally (*Kharif* 2017–2018)

Sl. no.	Crop name	No. of plots	Area (ha)	%
1	Cereals – *jowar*, maize, paddy	751	126.97	34.4
2	Pulses – beans, carom seeds, *tur*	77	17.23	4.7
3	Commercial crops – coriander and cotton	787	176.34	47.8
4	Vegetable crops – brinjal, cabbage, carrot, cauliflower, chilly, tomato	86	9.8	2.7
5	Flowers – carnation, rose	96	11.25	3.0
6	Fodder crops – fodder maize, Guinea grass, Napier grass	49	4	1.1
7	Fruit crops – guava, mango, orange	12	7.35	2.0
8	Plantation and spice Crops – teak, Malabar neem, palm tree, sandal tree	46	15.93	4.3
Total		1904	368.87	100

Table 8.8 Crops enumerated in Kothapally

Cropping pattern	Crop category	Crop name	No. of plots	Area (ha)	Total (ha)
Single crop	Cereals	*Jowar* (sorghum)	2	0.12	126.97
		Maize	423	99.22	
		Paddy	326	27.63	
	Pulses	Beans	5	0.47	17.23
		Carom seeds	1	0.19	
		Tur (pigeon pea)	71	16.57	
	Commercial crops	Coriander	1	0.1	176.34
		Cotton	786	176.24	
	Vegetable crops	Brinjal	4	0.64	9.8
		Cabbage	9	1.51	
		Carrot	6	0.67	
		Cauliflower	1	0.24	
		Chilly	9	1.04	
		Lady's finger	3	0.47	
		Tomato	54	5.23	
	Flowers	Carnation	18	2.15	11.25
		Rose	78	9.1	
	Fodder crops	Fodder maize	4	0.43	4
		Guinea grass	39	3.1	
		Napier grass	6	0.47	
	Fruit crops	Guava	1	0.35	7.35
		Mango	9	3.58	
		Orange	2	3.42	
Others	Plantation and spice crops	Teak	2	0.66	15.93
		Malabar neem	20	8.94	
		Palm tree	22	5.82	
		Sandal tree	2	0.51	
Total			1904	368.85	368.85

Table 8.9 Changing cropping pattern in Kothapally

| Land type | Crop category | Kharif crops – 2017 as per survey | | | Kharif crops – 2003–2004 (as per ICRISAT survey 2003–2004) | |
		Crop name	No. of plots	Area in ha	Crop name	Area in ha
Agricultural land	Cereals	Jowar	2	0.12	Sorghum	96.22
		Maize	423	99.22	Maize	78.85
		Paddy	326	27.63	Paddy	30.10
	Pulses	Beans	5	0.47	Beans	0.00
		Black gram	0	0.00	Black gram	0.81
		Carom seeds	1	0.19	Vamu	2.83
		Green gram	0	0.00	Green gram	0.81
		Tur	71	16.57	Pigeon pea	2.45
	Commercial crops	Coriander	1	0.10	Coriander	1.21
		Cotton	786	176.24	Cotton	78.28
		Sugarcane	0	0.00	Sugarcane	7.69
	Fruit crops	Guava	1	0.35	Guava	0.00
		Mango	9	3.58	Mango	0.00
		Orange	2	3.42	Orange	0.00
	Vegetable crops	Brinjal	4	0.64	Vegetables	18.56
		Cabbage	9	1.51		
		Carrot	6	0.67		
		Cauliflower	1	0.24		
		Chilly	9	1.04		
		Lady's finger	3	0.47		
		Tomato	54	5.23	Tomato	16.45
	Oil seeds	Sunflower	0	0.00	Sunflower	12.69
	Tubers	Onion	0	0.00	Onion	2.83
		Turmeric	0	0.00	Turmeric	10.08
	Flowers	Carnation	18	2.15	Flowers	2.23
		Rose	78	9.10		
	Fodder crops	Fodder maize	4	0.43	Fodder	0.51
		Guinea grass	39	3.10		
		Napier grass	6	0.47		
	Plantation and spice crops	Teak	2	0.66	Teak	0.00
		Malabar neem	20	8.94	Malabar neem	0.00
		Palm tree	22	5.82	Palm tree	0.00
		Sandal tree	2	0.51	Sandal tree	0.00

(continued)

Table 8.9 (continued)

Land type	Crop category	Crop name	No. of plots	Area in ha	Crop name	Area in ha
			Kharif crops – 2017 as per survey		Kharif crops – 2003–2004 (as per ICRISAT survey 2003–2004)	
	Fallow land	No crop (current fallow land)	202	28.58	No crop (current fallow land)	0.00
		Permanent fallow	174	60.16	Permanent fallow	0.00
Non-agricultural land	Non-agricultural land	Non-agricultural land (village area, built-up, road, stream, pond, road ROW)	242	46.96	Non-agricultural land (village area, built-up, road, stream, pond, road ROW)	0.00
	Total		2522	504.55	Total	362.62

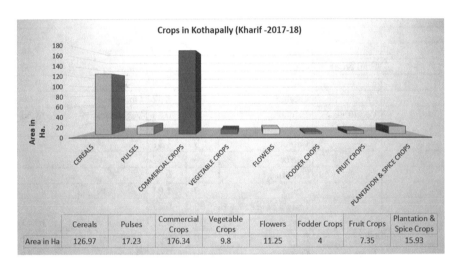

Fig. 8.17 Extent of crops in Kothapally (Kharif 2017–2018)

Enquiry with farmers about the type of seed used revealed that most of the crops are of hybrid varieties. Only few plots are with local varieties (particularly maize, paddy and *tur*). Almost all the plots with cotton are hybrid varieties except for few plots. Double cropping appears to be common in Kothapally wherever irrigation sources are developed. It was observed that many farmers have started sowing *rabi* (*post-rainy*) crops. Common observation was that most of the plots with maize, carrot and chilli during *kharif* are used for Bengal gram (chickpea). Plots with *tur* and cotton crops are not cultivated in the next season.

Current study focused on generation of crop information at plot level using geospatial technology. It has not only generated the crop data at plot level but has

Fig. 8.18 Crops grown in Kothapally

Fig. 8.19 Crops grown in Kothapally

developed a long-lasting spatial database of holdings. Repetition of crop data updation with this reference platform becomes easy and economical and provides
insights to farmers as well as GP. Farmers can analyse the data with respect to crop
performance within village and understand practices and inputs used for optimised
cultivation gaining better revenues.

Fig. 8.20 Crops grown in Kothapally

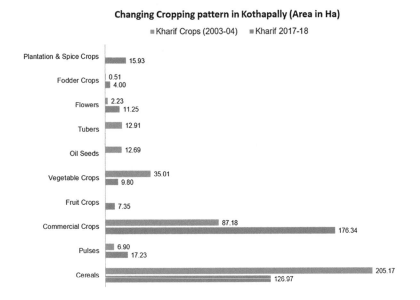

Fig. 8.21 Variation of kharif crops over a decade in Kothapally

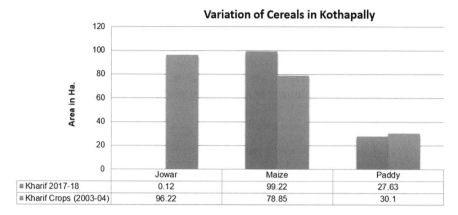

Fig. 8.22 Variation of cereals over a decade in Kothapally

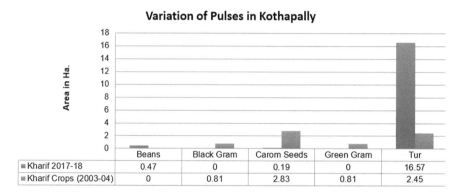

Fig. 8.23 Variation of pulses over a decade in Kothapally

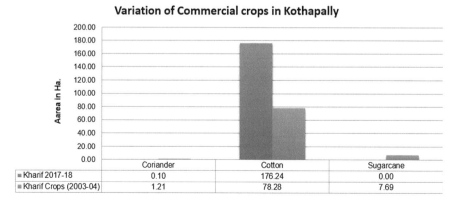

Fig. 8.24 Variation of commercial crops over a decade in Kothapally

The methodology adopted was simple. Any local farmer/youth who can operate smartphone will be able to update the information digitally. The updation of crops with smartphone takes less than 15 days for a season for a person who could be engaged by GP. If the process is repeated for all the three seasons, the data will be useful in understanding the season-wise extent of crops and yield. GP can plan for storage facilities and help farmers to find optimal market for their produce.

8.7   Irrigation

Geo-stamping approach for land-use and crop mapping was integrated with mapping sources of water and irrigation assets in Kothapally. The information not only indicates extent of development of resources but also provides insights of water conservation and management.

8.7.1   Sources of Irrigation

Rainfall is the only source of irrigation in Kothapally. In order to protect crops and to get better yield, few farmers have developed groundwater sources for cultivation. The survey during November 2017 indicated 109 borewells and 58 open wells (Table 8.10 and Fig. 8.25). During the survey, it was observed that few farmers pump the water from borewells to open wells also. In most of the cases, sprinkler and drip irrigation have been practised.

Spatial distribution of structures (Fig. 8.25) indicates that groundwater withdrawal is on the rise. Many of the open wells are becoming unusable. The defunct open wells (Fig. 8.26) are located in the vicinity of the borewells indicating that water table is being lowered due to pumping from deeper aquifers. Even few borewell shaves become defunct (Fig. 8.27 and Table 8.11).

Table 8.10 Sources of irrigation in Kothapally

| ICRISAT report 10 | | | | | As per the survey during November 2017 | | |
| Structures | | | Shankarpally GP report | | | | |
Borewells before project initiation	Borewells drilled during the project period (1999–2001)	Open wells	No. of borewells	Powered for agriculture	Borewells	Open wells	Remarks
15	55	62	176	75	2	–	Drilled
					97	21	Functioning
					10*	37	Defunct
					109	58	

*Out of ten defunct borewells, five have gone dry (pump removed), three dried up (pump still retained) and two reduced yield and not being used

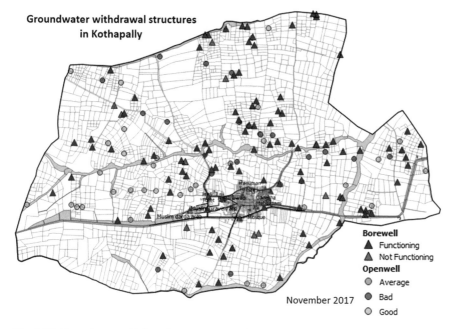

Fig. 8.25 Groundwater withdrawal structures in Kothapally

Fig. 8.26 Unused open wells in Kothapally

Fig. 8.27 Non-functioning borewells in Kothapally

Old records (Sreedevei et al. 2004) indicate that there were only 70 borewells and 62 open wells. The Shankarpally GP report provides the number as 176 and only 75 are energised for agriculture. It is not clear whether the information pertains to only Kothapally. Due to drilling deeper borewells, the water table has been lowered and the yield from the open wells was insufficient. Therefore, farmers must have stopped using open wells directly for irrigation. As such, 37 open wells are serving currently as rainwater recharge/harnessing structures.

Information from the farmers indicates (Table 8.12) that in the recent times, farmers are going for deeper borewells as most of the wells shallower than 300 ft are showing decline in the yield or not yielding for continuous pumping.

8.7.2 Rainwater Harnessing Structures

Rainwater harnessing appears to be one of the important interventions of the watershed management programme. Masonry check dams, recharge pits and sunken pits (Fig. 8.28) are observed at many locations. Location of the structures is shown in Fig. 8.29. Rainwater is being collected, and it percolates into the shallow aquifers and builds up the water table. Farmers have opined that after these structures were built, substantial improvements in the water levels in open wells have been observed. As a result, these farmers have been able to tap groundwater for irrigation in the areas near to the open wells. The records (posters displayed in Kothapally) indicate more rainwater harnessing structures (Table 8.13). Field survey team could not locate all these structures as they might have been concealed under the vegetation and/or standing crop.

Table 8.11 Details of defunct and low-yielding borewells in Kothapally

Sl	Latitude	Longitude	[b]Year of drilling	Yield[a] (inches) when drilled	Depth in feet	Depth of casing in feet	Status	Remarks
1	17.37793216	78.11457246	2007	2	60	20	Functioning	No water in summer season
2	17.38261576	78.12167286	1997	2	180	50	Functioning	Insufficient water
3	17.37798955	78.11429543	2012	2	200	45	Functioning	Insufficient water
4	17.38752769	78.11909232	2012	2	200	85	Functioning	Insufficient water
5	17.38756688	78.11868884	2012	2	250	70	Functioning	Not used in kharif season due to water shortage
6	17.3816213	78.12155854	1997	2	300	40	Functioning	Goes dry during summer
7	17.37128284	78.11302958	2005	2	350	25	Functioning	Low and intermittent yield since 2014
8	17.38851412	78.12403117	2006	2	150	20	Not functioning	Borewell dried
9	17.37652246	78.12175986	2012	2	200	40	Not functioning	Borewell dried
10	17.38180255	78.10911273	2002	1	200	80	Not functioning	Very low yield (not using)
11	17.38383124	78.11753603	2009	1	250	80	Not functioning	Borewell dried
12	17.38568721	78.1261454	2005	1	250	80	Not functioning	Borewell dried
13	17.37794164	78.11456929	2016		300	120	Not functioning	Borewell dried
14	17.38116764	78.10968234	2013		300	100	Not functioning	Borewell dried
15	17.37299873	78.12022355	2012	1	300	80	Not functioning	Water shortage, not using
16	17.37286952	78.1198745	2007	1	340	80	Not functioning	Borewell dried
17	17.37572811	78.12524155	2003		350	80	Not functioning	Borewell dried

[a]Yield is expressed in inches by farmers. Generally, 1 in. corresponds to about 600 l per hour, and if it is 2 in., the discharge is around 3400 l per hour; 2 in. is considered as high yield
[b]Year of drilling is the answer provided by farmer. The year may not be accurate as it is from the memory; the response was given by the farmer

Table 8.12 Depth range of borewells in Kothapally

Sl. no.	Depth range (feet)	Number of borewells
1	Less than 200	31
2	200 to 300	42
3	300 to 400	26
4	Above 400	10
Total		109

Masonry Check dam   Recharge well

Sunken Pit   Percolation Pit

Fig. 8.28 Rainwater harnessing structures in Kothapally

8.7.3 Area Under Irrigation

The Kothapally Village does not come under any canal irrigation/tank irrigation command area. Farmers have resorted to use of groundwater for either protecting crops during deficit rains or many a times growing crops in post-rainy seasons. Farmers are in a position to irrigate substantial extent of lands (Fig. 8.30). It is observed that 109.49 ha (Table 8.14) are under irrigation. Farmers are adopting sprinkle/drip irrigation method to optimise the use of water.

The spatial distribution of irrigated plots (Fig. 8.30) and the sources of irrigation (particularly the borewells) indicate that groundwater is the sole source. It may also be noted that borewells are getting defunct due to low yield, and most of the open wells are either used for storing the pumped water from borewells or storing the rainwater. The density of borewells (Fig. 8.30) may be kept in mind for further development of groundwater.

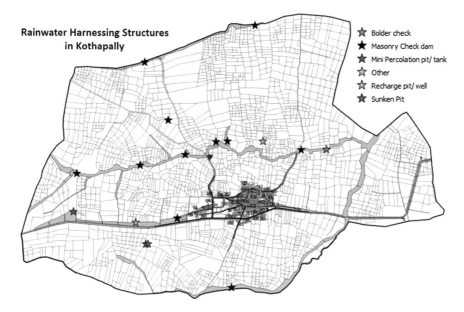

Fig. 8.29 Location of rainwater harnessing structures in Kothapally Village

Table 8.13 Types of rainwater harnessing structures in Kothapally Village

Sl. no.	Structure type	No. of structures geo-stamped	As per the poster in Kothapally
1	Boulder check/gully plugs	1	97
2	Masonry check dam	11	11
3	Mini percolation pit/tank	2	47
4	Other structures/gabion structure	2	1
5	Recharge pit/well	1	39
6	Sunken pit	1	51
Total		18	246

Filed inventory was conducted during November 2017. Most of the boulder checks/gully plugs were not visible due to siltation and growth of weeds. Also, there are few more check dams located within the watershed but outside the limit of revenue boundary of the village

An exercise on relationship of location of borewells with holdings (Table 8.15 and Fig. 8.31) was done to understand the efforts made by different farmers in developing sources of irrigation.

The water assets distribution with respect to agricultural holdings indicates that more than 60% of the borewells are drilled by the marginal farmers followed by small farmers and more than 80% of the groundwater assets are developed by these marginal and small farmers. Ensuring water (deficit from rainfall) to the crops appears to have been important consideration for all these farmers.

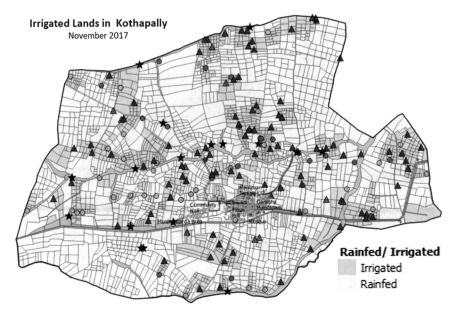

Fig. 8.30 Irrigated areas in Kothapally Village

Table 8.14 Extent of irrigation in Kothapally Village (November 2017)

Sl. no.	Source of water	GP records Area (ha.)	Survey during November 2017 Area (ha.)	No. of plots
1	Rainfall	210.4	348.1	1346
2	Borewell	150	109.49	934
3	Check dams (watershed)	63	–	–

Table 8.15 Distribution of water assets vs. land holdings in Kothapally

Sl. no.	Land holding types	No. of holdings	Area (ha)	No. of borewells	No. of open wells
1	Marginal holdings (<1 ha)	399	176.24	64	26
2	Small holdings (1–2 ha)	87	120.76	30	19
3	Semi-medium holdings (2.1–4 ha)	28	80.61	10	8
4	Medium holdings (4.1–10 ha)	7	33.45	3	3
5	Large holdings (>10 ha)	3	38.95	2	2
6	Non-agricultural land	–	43.34	–	–
7	Govt. land	–	11.20	–	–
	Total	524	504.55	109	58

Fig. 8.31 Distribution of water assets in relation to land holdings in Kothapally

Water resources are very important input to agriculture, and at the same time judicial use needs to be kept in mind. Inventory of groundwater assets and pattern of dwindling yield in borewells need to be regularly monitored so that better recharging – both natural and augmented (similar to the interventions made as a part of watershed management programme) – could be put in place to protect the groundwater. The data on borewells and irrigation is useful to local farmers and particularly the local watershed management committee who can encourage farmers to optimise the use and adopt efficient means of irrigation. It is advantageous to update the information at least once a year on such sources and take step towards further development/regulation.

8.8 Household Pattern

8.8.1 *Household Survey*

Each dwelling unit in the Kothapally settlement was mapped using the high-resolution satellite image, and each unit was provided with a unique number (Fig. 8.32). A large-scale map was printed with unique house number and provided to enumerator during the survey. A separate application was developed for recording the information (Fig. 8.7). During the survey, the enumerator visited each

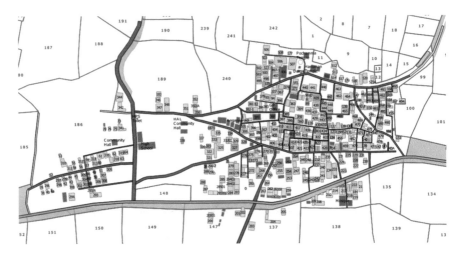

Fig. 8.32 Kothapally settlement with dwelling units and roads

household and recorded the attribute information along with key information of the land holding (RTC) so that crop plots can be correlated with land holding data.

8.8.2 Habitat Status

In addition to the crop inventory, a detailed survey of the Kothapally habitat was also conducted mainly to understand the status of settlement. Information related to house types and other infrastructures are provided here. Kothapally is developed on a local water divide and on the left side of the road leading to Hyderabad. It is a small settlement with as many as 452 dwelling units. The internal roads within the settlement (Fig. 8.33) are in good condition, and most of them are made of cement concrete.

The Kothapally Village appears to be having all the required basic infrastructural facilities, and even there are wastewater treatment facilities (which were built as part of watershed management interventions). Almost all the houses have sanitary facilities and are connected with underground drainage.

There is considerable variation in the type of roofs of houses in Kothapally. The roof types indicate that almost all the houses are neatly built and are properly covered. The roof material varies within the village (Fig. 8.34 and Table 8.16). The GP record might have included the houses in a satellite habitat which is outside the revenue limits of Kothapally.

All the dwelling units are used for residential purposes besides the structures for school and non-residential uses like shops, temple, etc. Occupation of the people residing also has been collected and presented in Table 8.17 and Fig. 8.35. The distribution of agriculturists by holding type is shown in Fig. 8.36.

The data indicates that most of the residents of Kothapally are farmers with land and a substantial number of families residing in the village are landless labourers.

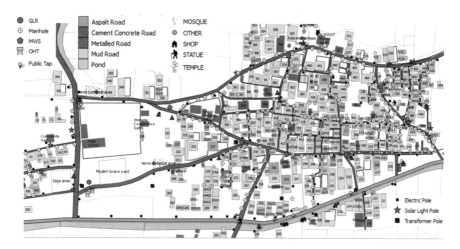

Fig. 8.33 Kothapally habitat with major infrastructures

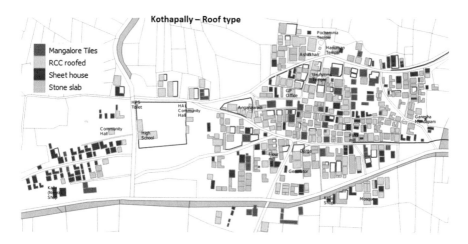

Fig. 8.34 Roof types of the buildings in Kothapally

Table 8.16 Type of houses in Kothapally

Sl. no.	Type of house	As per survey	As per GP records
1	Mangalore tiles	42	184
2	RCC roof	115	126
3	Sheet house (asbestos)	116	143
4	Stone slab	179	0
5	Thatched	0	49
Total		452	502

Table 8.17 Occupancy type in Kothapally

Occupation type	No. of houses
Agriculturist	266
Employee	7
Landless labourer	108
Self-employed	12
No details	59
Total	452

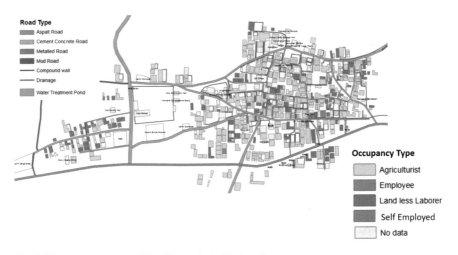

Fig. 8.35 Occupancy type of dwelling units in Kothapally

Fig. 8.36 Distribution of farming households in Kothapally

Table 8.18 Land values in Kothapally Village

Land value within the settlement (Rate in Rs. Per Sq.Ft)			
GP records		Enquiry during survey	
Only land	Building	Only land	Building
33.3	405	56–93	464–743
Agricultural land (enquiry during survey): rate in Rs. per acre			
Near to road		Other lands (deeper in the village)	
Rs. 3 to six million		Rs. 1 to two million	

8.8.3 Property/Land Prices

General enquiry was made with people in the village, and information related to land prices was collected from the gram panchayat office. The price details are presented in Table 8.18.

Land values appear to be on the higher side. It is observed and people say that many investors are moving in to villages to buy lands. Lands located on the road fetch very high value against those deeper inside the village. Since the village is located near to Hyderabad and major highways also provide connectivity, lands attract good value.

8.9 Summary

1. Land-use analysis, cropping pattern and sources of irrigation are important for planning and improving the rural economic aspects. In India, as villages are defined with a clear administrative boundary and land holdings forming a specific ecosystem with finite, varying resources, study of land use assumes significance. There is no internationally accepted land-use classification mainly because the classes have been developed for a specific purpose and scale. India has adopted ninefold classification which is being used for most of the planning purpose. Different methods are in use for deriving and analysing the land-use information across the world, and for generating the same at micro-level, the geospatial technology developed and tested by ICRISAT has been found to be useful and adopted.

2. Kothapally Village covers the Adarsha watershed wherein ICRISAT has adopted a participatory consortium approach with emphasis on harnessing rainwater besides enhancing and sustaining productivity in soils and adoption of improved crop management practices. ICRISAT has been documenting the results of the interventions over the years in the watershed and has embarked up on plotwise crop inventory in the entire Kothapally Village along with household survey during *kharif* season of 2017–2018. Geo-stamping approach has been adopted for generating the spatial data in Kothapally Village. The data has been organised in QGIS.

3. The village has 524 holdings of varying sizes. The average holding is 0.96 ha. Kothapally is characterised by a large number of marginal holdings, i.e. 399 holdings with less than 1 ha. Kothapally GP needs to update and maintain records as per the land holding classification. The average land holding of Kothapally is lesser than that in Telangana (1.12 ha).

4. Land use in Kothapally is characterised by agriculture, and the land put to non-agricultural use is 9.31%, which covers the settlement and area used for roads and drainage courses. The GP records have shown a smaller area under this category; probably the settlement area was not properly determined, *or* it must have taken the area under "Gramthan" – the original area of settlement as per the revenue records. The village has substantial extent of permanent fallows (11.92%), and seasonal fallow accounts to 5.67%. The net sown area of 368.87 ha (73.1%) comprises of *kharif* and perennial crops. Water features such as streams, open wells, ponds and tanks covers 18.06 ha. Reasons for permanent fallow have been understood as follows:

   • Farmers who have migrated to urban area
   • Road side land being sold to commercial purpose
   • Government land distributed to poor not being cultivated
   • Underutilised government land
   • Other socioeconomic reasons

   It is observed that the semi-medium and medium holdings have been left as permanent fallows compared to other types of holdings.

5. Kothapally is characterised by unique cropping systems comprising of cereals, pulses, fruits, commercial crops, vegetables, flowers and plantations. Cotton is the major commercial crop and occupies an area of 176.24 ha. Paddy, maize and *jowar* are the important cereal crops together occupying an extent of 126.97 ha. Out of these, maize occupies an area of 99.22 ha. *Tur* (pigeon pea) occupies the most area (16.57 Ha.) amongst the pulses. Different varieties of vegetables are grown in Kothapally, and tomato covers larger area of 5.23 ha. Carnation and rose are the two important flower crops with rose covering an area of 9.1 ha. Fodder crops occupy an area of 4 ha, and fruits are grown in an area of 7.35 ha. Mango and orange are almost equally distributed. Plantations occupy an area of 15.93 ha., and Malabar neem occupies an area of 8.94 ha.

6. Comparison of the earlier data and current survey results indicate that the extent of cereals has remained almost the same, with sorghum being almost not cultivated. Amongst the pulses, *tur* (pigeon pea) appears to be gaining prominence and area under cotton has doubled. Sugarcane has disappeared.

7. Farmers have mentioned that they use hybrid seeds and only for few plots they use local varieties (particularly maize, paddy and tur). Almost all the plots with cotton are hybrid varieties except for a few.

8. Double cropping appears to be common in Kothapally wherever irrigation sources are developed. Common observation was that most of the plots with

maize, carrot and chilli during *kharif* are used for Bengal gram chickpea. Plots with tur and cotton crops are not cultivated in the next season.

9. Rainfall is the only source of irrigation in Kothapally. The village does not come under any canal irrigation/tank irrigation command area. Groundwater is being used for irrigating 109.49 ha. In order to protect crops and to get better yield, few farmers have developed groundwater sources for cultivation. The survey during November 2017 indicated 109 borewells and 58 open wells. Most of the farmers practise sprinkler and drip irrigation method.

10. Groundwater withdrawal is on the rise, and shallow open wells are becoming unusable. Farmers in the recent years are going for deeper borewells as most of the wells shallower than 300 ft are showing reduction in the yield or not yielding for continuous pumping. The water table is being lowered due to pumping from deeper aquifers. Seventy-eight borewells (72%) are having depth of 200 and above feet out of 109 existing borewells. Even few borewells have become defunct. As such, 37 open wells are serving currently as rainwater recharge/harnessing structures.

11. Masonry check dams, recharge pits and sunken pits constructed during the implementation of watershed interventions have been inducing recharge, and farmers have opined that substantial improvements in the water levels in open wells have been observed. As a result of this, farmers have been able to tap groundwater for irrigation in the areas near to the open wells.

12. Kothapally is a small settlement with as many as 452 dwelling units. The internal roads within the settlement are in good condition, and most of them are made of cement concrete. The village has all the required basic infrastructural facilities, and even there are wastewater treatment facilities (which were built as part of watershed management interventions). Almost all the houses have sanitary facilities and are connected with underground drainage.

13. There is considerable variation in the type of roofs of houses in Kothapally. The roof types indicate that almost all the houses are neatly built and properly covered. The roof material varies within the village. Forty per cent of houses are having stone slab roof, 26% with asbestos sheet roof, 25% with RCC roof and 9% with *Mangaluru* tile roof.

14. The data indicates that most of the residents of Kothapally are farmers with land (59%) and a substantial number of families residing in the village are landless labourers (24%).

15. It is observed and people opine that many investors are moving in to villages to buy lands, since the village is located near to Hyderabad and major highways provide connectivity. Value of open land is Rs. 33/ft^2 and with building is Rs. 405/ft^2 as per GP records within the settlement. Lands located on the road fetch very high value (Rs. 30–60 *lakhs*/acre) against those deeper inside the village (Rs. 10–20 *lakhs*/acre).

References

James R. Anderson, Ernest E. Hardy, John T. Roach, and Richard E. Witmer (1976). Geological Survey Professional Paper 964, A revision of the land use classification system as presented in U.S. Geological Survey Circular 671.

Barnes, C. P. (1936). *Land classification: Objectives and requirements*, U. S. Resettlement Administration and Utilization Division, Land Use Planning Publication I, pp. 35–36.

Bernetti, I., & Marinelli, N. (2010). Evaluation of landscape impacts and land use change: A Tuscan case study for CAP reform scenarios. *AESTIMUM, 56*(Giugno), 1–29.

Bernetti I., Franciosi, C., & Lombardi, G. (2006). Land use change and the multifunctional role of agriculture: A spatial prediction model in an Italian rural area. *International Journal* of Agrarian *Research, 5*(2/3).

Castelluccio, M., Poggi, G., Sansone, C., Verdoliva, L. (2015). *Land use classification in remote sensing images by convolutional neural networks*. arXiv:1508.00092v1 [cs.CV] 1 Aug 2015.

CEC [Commission of the European Communities]. (1993). *CORINE land cover – Guide technique*. Brussels.

Chattopadhyay, S., Krishna Kumar, P., & Rajalekshmi, K. (1999). *Panchayat resource mapping to panchayat – Level planning in Kerala: An analytical study. Kerala research Programme for local level development*. Thiruvanthapuram: Centre for Development Studies.

Dadhwal, V. K., Singh, R. P., Dutta, S., & Parihar, J. S. (2002). Remote sensing based crop inventory: A review of Indian experience. *Tropical Ecology, 43*(1), 107–122. 2002 ISSN 0564-3295.

De Wit, P., & Verheye, W. (2009). *Land use planning for sustainable development. Land use, land cover and soil sciences* (Vol. III). Encyclopedia of Soils in the Environment.

Di Gregorio, A., & Jansen, L. J. M. (2000). *Land cover classification system (LCCS): Classification concepts and user manual*. Rome: FAO.

Eiten, G. (1968). *Vegetation forms. A classification of stands of vegetation based on structure, growth form of the components, and vegetative periodicity* (Boletim do Instituto de Botanica (San Paulo), No. 4).

FAO. (1976). A framework of land evaluation. *FAO Soils Bulletin, 32*, Rome, p. 77.

Fosberg, F. R. (1961). A classification of vegetation for general purposes. *Tropical Ecology, 2*, 1–28.

Jadab Chandra Halder (2013). Land use/land cover and change detection mapping in Binpur-II block, Paschim Medinipur District, West Bengal: A remote sensing and GIS perspective. IOSR Journal of Humanities and Social Science (IOSR-JHSS) Volume 8, Issue 5, March–April (2013)*, PP* 20–31.

Hettig, E., Lay, J., & Sipangule, K. (2016). Drivers of households' land-use decisions: A critical review of micro-level studies in tropical regions. *Land*, (5), 32. https://doi.org/10.3390/land5040032.

ICRISAT. (2004). *Watershed management and livelihood options*, survey data.

Kannadasan, K., Ganesh, A., & Vinothkanna, S. (2017). An analysis of nine fold classification land use land cover changes In Tiruvannamalai District, Tamil Nadu, India. *International Education & Research Journal [IERJ]* E-ISSN No: 2454-9916|Volume: 3|Issue: 6|June 2017.

Karlsson, I., & Rydén, L. (2012). *Rural development and land use*. ISBN 978-91-86189-11-2.

Kristina Nilsson, L., & Rydén, L. (2012). *Spatial planning and management, in rural development and land use* (I. Karlsson and L. Rydén, Eds.). ISBN 978-91-86189-11-2.

Kuechler, A. W., & Zonneveld, I. S. (Eds). (1988). *Vegetation mapping* (Handbook of vegetation science, Vol. 10). Dordrecht: Kluwer Academic.

LaGro, J. A. (2005). *Land-use classification*. Encyclopedia of Soils in the Environment.

Lambin, E. F., Geist, H. J., & Lepers, E. (2003). Dynamics of land-use and land cover change in tropical regions. *Annual Review of Environment and Resources, 28*, 205–241.

Mandal, R. B. (1990). *Land utilization: Theory and practice* (pp. 24–41). New Delhi: Concept Pub. 130-169.

Marsh, W. M. (1991). *Landscape planning: Environmental applications* (2nd ed.). New York: Wiley.

Miller, D., Matthews, K., Buchan, K.. (2012). Determining the land use and farm-type mixes of land capability for agriculture class groupings in the regions of Scotland and the characteristics of holdings containing both land capability for agriculture classes 1 to 5.3 and 6.1 to 7 land. The James Hutton Institute.

Mueller-Dombois, D., & Ellenberg, J. H. (1974). *Aims and methods of vegetation ecology.* New York/London: Wiley.

Naik, G., Basavaraj, K. P., Hegde, V. R., Paidi, V., & Subramanian, A. (2013). Using geospatial technology to strengthen data systems in developing countries: The case of agricultural statistics in India. *Applied Geography, 43*(2013), 99–112.

Petja, B., Nesamvuni, E., & Nkoana, A. (2014). Using geospatial information technology for rural agricultural development planning in the Nebo Plateau. *South Africa: Journal of Agricultural Science, 6*(4), 2014.

PIB. (2015). Press Information Bureau Government of India, Ministry of Agriculture & Farmers Welfare; Highlights of Agriculture Census 2010–11.

Premakumar, K., & Vinothkanna, S. (2015). Spatio-temporal analysis of land use in Palakkad District, Kerala. *International Journal of Current Research, 7*(11) (ISSN: 0975-833X).

Richards, J. F., & Flint, E. P. (1994). *Historic land use and carbon estimates for South and Southeast Asia: 1880–1980.* https://doi.org/10.3334/CDIA/lue.ndp046.

Roy, P. S., & Giriraj, A. (2008). Land use and land cover analysis in Indian context. *Journal of Applied Sciences, 8*, 1346–1353.

SD21. (2012). United Nations Department of Economic and Social Affairs Division for Sustainable Development.

Sen, P. (2016). Shaping the fringes for the expansion of cities-a study on Hoskote in the greater Bengaluru urban area. *Transactions, 38*(1), 2016.

Sokal, R. (1974). Classification: Purposes, principles, progress, prospects. *Science, 185*(4157), 111–123.

Sombroek, W. (1992). *Land use planning and productive capacity assessment*, FAO.

Sreedevi, T. K., Shiferaw, B., & Wani, S. P. (2004). *Adarsha watershed in Kothapally: Understanding the drivers of higher impact* (Global theme on agroecosystems report no. 10). Patancheru 502 324, Andhra Pradesh, India: International Crops Research Institute for the Semi-Arid Tropics. 24 pp.

Stamp, L. D. (1951). *Land for tomorrow: The under-developed world*, Bloomingtom, 1951.

Suresh Kumar, S. (2014). *Land accounting in India: Issues and concerns.* https://unstats.un.org/unsd/envaccounting/seeales/egm/LandAcctIndia.pdf. Contribution to System of Environmental-Economic Accounting 2012—Experimental Ecosystem Accounting. United Nations.

Tian, H., Banger, K., Bo, T., & Dadhwal, V. K. (2014). History of land use in India during 1880–2010: Large-scale land transformations reconstructed from satellite data and historical archives. *Global and Planetary Change, 121*(2014), 78–88.

Tiwari, R., Indu, K. M., Killi, J., Kandula, K., Bhat, P., Ramjee Nagarajan, R., Kommu, V., Kameshwar, R. K., & Ravindranath, N. H. (2010). Land use dynamics in select village ecosystems of southern India: Drivers and implications. *Journal of Land Use Science, 5*(3), 197–215. https://doi.org/10.1080/1747423X.2010.500683.

UNESCO. (1973). *International classification and mapping of vegetation.* Paris.

Chapter 9
Mainstreaming of Women in Watersheds Is Must for Enhancing Family Income

Girish Chander, S. P. Wani, D. S. Prasad Rao, R. R. Sudi, and C. S. Rao

Abstract Despite the fact that women are the world's principal food producers and providers, they have long been deprived of their due share and identity. Kothapally is one of the initial watershed projects that demonstrated on ground that a holistic development model not only conserves natural resources for sustainable productivity and income improvement but also harnesses the synergies to tailor the benefits in mainstreaming women farmers. This has showcased the model to focus on selective activities that directly benefit women. Some important activities that increase incomes of women revolve around interventions like milk production, kitchen gardens, composting, value addition, non-farm livelihoods through capacity building, collectivization and market linkages.

Keywords Feed and fodder · Food and nutrition security · Kitchen gardens · Livestock-based livelihoods · Spent-malt · Waste recycling

9.1 Kothapally Watershed: An Inclusive Holistic Approach for Mainstreaming Women

In spite of women's substantial contribution to agricultural production, women in general practice are marginalized when it comes to land ownership, decision-making and their share of income. Earlier watershed guidelines covered land-based activities only, and as a result in India, most women who have no land rights were not direct beneficiaries of the watershed program. As a result, as indicated earlier in Chap. 2, without any tangible economic benefits, 50% of the

G. Chander (✉) · D. S. Prasad Rao · R. R. Sudi · C. S. Rao
ICRISAT Development Centre, Research Program Asia, International Crops Research Institute for the Semi-Arid Tropics (ICRISAT), Hyderabad, Telangana, India
e-mail: g.chander@cgiar.org

S. P. Wani
Former Director, Research Program Asia and ICRISAT Development Centre, International Crops Research Institute for the Semi-Arid Tropics (ICRISAT), Hyderabad, Telangana, India

© Springer Nature Switzerland AG 2020
S. P. Wani, K. V. Raju (eds.), *Community and Climate Resilience in the Semi-Arid Tropics*, https://doi.org/10.1007/978-3-030-29918-7_9

population did not participate actively in the watershed development activities. Alongside ongoing policy and social reforms, it is high time that we make concerted efforts within existing opportunities for empowering women farmers by providing income enhancement options. The integrated watershed development approach framework followed in Kothapally watershed is not only one of the tested, sustainable and eco-friendly approaches to conserve soil and water resources for enabling productivity improvement and diversification but also mainstreaming of women through improving livelihoods and addressing equity issues (Wani et al. 2012, 2014a). By addressing the core requirement of providing knowledge about technical and financial aspects of income-generating enterprises along with a hand-holding support, women can carve their way out to improve incomes and secure justified identity in the society; this is an exemplar case study in Kothapally. The holistic approach adopted in Kothapally was to focus on strengthening certain women-centric enterprises along with collectivization of participating women for addressing risks and effective market linkages. The focus of prominent enterprises were milk production through addressing feed, fodder and breed improvement, value addition, composting, nutri-kitchen gardens, seed banks and petty shops in the village.

9.2   Strengthening Livestock-Based Enterprises

Livestock are integral part of farming, and especially dairy and poultry are in general in domain of women for meeting the household nutritional requirements as well as marginal sources of income through sale of extra production by the women farmers. Initial baseline surveys and farmer interactions showed very low milk productivity level in the watershed mainly attributed due to scarcity and poor quality fodder and low-yielding animals. With the increased water availability in the watershed, not only crop productivity was increased but also cropping intensity along with crop diversity resulting in additional fodder availability in the Kothapally watershed. With the increasing demand by the growing population, higher incomes and more health consciousness, there was rising demand for milk. Based on the increased fodder availability and also milk demand, this activity was prioritized for women farmers' income addition in the watershed. The baseline survey had indicated that there was no marketable surplus for milk in the village in 1998–1999 as milk production in the village was only around 250 l per day which was consumed locally (Shiferaw et al. 2002; Wani and Shiferaw 2005). Livestock is one of the most important sources of income for the families in dryland areas. Generally, women members in the family take care of the livestock-based activities, and it also results in improving family incomes. Soon after the watershed activities were initiated, the first and foremost thing was to improve the animal breed along with fodder development and subsequently introducing feed/concentrates for the livestock in the watershed to benefit the women farmers.

9.2.1   Expansion and Breed Improvement

During baseline, one of the major reasons identified for low productivity of live-stock was prevalence of low-yielding (1–3 l per animal per day) animals in the village. Therefore, in partnership with BAIF NGO partner, breed improvement was taken as a prioritized activity in the watershed. To address the issue of improving productivity of animals in the village, we adopted artificial insemination (AI) approach using the semen of improved cow breed Holstein, and for buffaloes we used Jafarabadi breed semen. Initially to establish the AI Centre, there was resistance from the department of Animal Husbandry of the Government of Andhra Pradesh. The government AI Centre was located at Shankarpally which is about 13 km away from the Kothapally. We took up the issue with the district Collector highlighting that when animals are in heat it is not feasible to take animals 30 km away up and down. Moreover, there was no regularity of availability of the expert or the semen. With the approval from the district Collector, the Watershed Committee received the permission from the GoAP to start AI Centre in the village. We brought Bhartiya Agro Industries Foundation (BAIF) as a consortium partner in the Watershed Consortium. The BAIF had proven expertise for AI in the country, and we had good experience working with BAIF (Pune, Maharashtra) in the Madhya Pradesh Milli Watershed. We developed a model with BAIF and requested them to have a sustainable model to ensure that within 5 years the model should become self-sustainable. For this the village youth were trained in AI process, and villagers had to just give a phone call to the AI person about the right stage of the animal to conceive, and the representative will reach the farmer's home along with mobile AI unit. In consultation with the Watershed Committee, a subsidized fee per AI was fixed (Rs 30 animal^{-1} up to 2004, Rs 50 animal^{-1} up to 2014 and Rs 100 animal^{-1} from 2015 onwards). The semen and the needed instruments including motorcycle, nitrogen gas container and phone were provided to the AI person. Following this approach in addition to Kothapally watershed, other four satellite watershed villages were also included in the AI program. Following this approach during the period 2003 till 2017, a total of about 4500 artificial inseminations (AI) were done in the local livestock – ~3000 in buffaloes and ~1500 in cows. As a result, around 1500 AIs were confirmed and improved breed calves were born. The productivity of animals due to AI increased and the F_1 animals were yielding 2–5 l per day as compared to 1–3 l per animal per day. In 2013, there was marketable surplus milk production with 2100 l per day.

9.2.2   Fodder Development

With the increased water availability, increased crop productivity quantity of crop residues also increased (Fig. 9.1). In addition, introduction of improved dual purpose cultivars improved quality fodder also. With water availability, farmers started cultivating green fodder in the watershed (Fig. 9.2).

Fig. 9.1 Good crop growth and biomass (which is used as cattle fodder) with improved practice in Kothapally watershed

Fig. 9.2 Improved fodder grass (Napier) cultivation in Kothapally watershed

It is well established fact that for good health and higher productivity, dairy cattle need good quality green roughages; more precisely dry matter requirement is around 2–3% of body weight. Quality grasses (e.g. Guinea, Napier, etc.) and legume fodder (e.g. cowpea, Lucerne, etc.) are required to meet protein and other nutrient requirements. In general, green fodder and legume component are lacking or in deficit, and straw is mostly major roughage. With the objective of boosting livestock productivity, green fodder production is promoted.

Moreover, the outcome of soil degradation in predominant crop-livestock farming system in the drylands is far beyond reducing grain production and quality; it also affects livestock feed quantity and quality (Blümmel et al. 2009; Haileslassie et al. 2011). In view of the increasingly important role of crop residue as feed components, the effects of soil health building through nutrient balancing on feed availability and feed quality are very important and show up in potential milk yield per ha by as high as 40% (Haileslassie et al. 2013). The role of soil health building in enhancing food quantity and quality and helping individuals and communities to build sustainable food security is well demonstrated in Kothapally.

9.2.3 Reliance for Procurement of Milk in the Village

As Kothapally was not on the Cooperative Milk Dairy procurement route and in 1998–1999, there was no marketable surplus of milk in the village, and there was no established milk marketing arrangement for the village. As the milk production started increasing, farmers started looking for the marketing channels. Initially one or two farmers started pooling milk together to take it to city for marketing. However, soon it became evident that to ensure good benefits to the farmers, we have to eliminate middlemen in the marketing channel. In discussions with the Reliance Industries – a supermarket chain, was initiated, and in 2007an automated milk procurement centre was established in the village. This centre provided a fair and fixed price with computerized fat % estimation and weighing facility for the farmers. In addition, the centre also provided animal feed to the farmers on credit basis as the milk price was paid every fortnightly basis. With this direct marketing to the Reliance, automatically more and more farmers started selling milk in the village itself. The details of the milk rates as well as the quantity of milk procured yearly basis are indicated in Table 9.1.

9.2.4 Spent Malt as a Microenterprise to Benefit Women and Milk Producers: A Business Model

Productivity of milch animals and business profitability is largely dependent on fodder/feed availability as well as its cost and quality. In a common situation of lack of green fodder in general, especially with lactating animal, feed/concentrate is required to make up for lacking protein and nutrients. Spent malt is a good feed material for livestock for improving health, milk yield and fat content. Spent malt is a byproduct of brewing industry, consisting of the residue of malt and grain which contains carbohydrates, proteins, lignin and water-soluble vitamins as animal feed.

Table 9.1 Total milk sale and price realized in Kothapally village

Year	Milk collection by Reliance/ Heritage group (litre)	Milk collection by Rajarajeswari group (litre)	Collection by private vendors (litre)	Total milk sale in the village (litre)	Cost of milk based on fat content (Rs/ litre)		
					5% fat	7% fat	11% fat
2006	–	–	95,000	95,000			
2007	90,000	–	135,000	225,000	14.0	22.4	28.0
2008	72,000	–	222,000	294,000	16.0	22.4	32.0
2009	180,000	–	252,000	432,000	17.5	24.5	35.0
2010	144,000	–	257,000	401,000	20.0	28.0	40.5
2011	86,000	28,800	272,000	386,800	22.5	31.5	45.0
2012	82,600	43,200	228,500	354,300	25.0	33.0	50.0
2013	85,000	50,400	235,500	370,900	25.0	35.0	50.0
2014	80,200	57,600	246,000	383,800	27.5	38.5	55.0
2015	75,300	64,800	240,500	380,600	31.5	44.5	63.0
2016	57,600	64,800	229,500	351,900	29.0	40.5	58.0
2017	65,200	60,200	227,000	352,400	30.0	42.0	60.0

Table 9.2 Nutritive value of spent malt and recommended mineral mixture

Nutrient	Spent malt: nutrient composition	2 kg spent malt: nutritive value	100 g mineral mixture: nutritive value
Nitrogen (%)	3.66%	–	–
Protein (%)	22%	440 g	–
Phosphorus	0.46%	9.20 g	9.00 g
Iron	205 ppm	0.41 g	0.40 g
Zinc	52 ppm	0.11 g	0.30 g
Copper	248 ppm	0.50 g	0.06 g
Manganese	29.5 ppm	0.06 g	0.10 g
Sulphur	2655 ppm	5.31 g	0.40 g
Calcium	2098 ppm	4.20 g	18.0 g
Magnesium	1602 ppm	3.21 g	5.00 g

It is quite palatable and is readily consumed by animals. Two kilograms of spent malt (on dry weight basis) provide about 400 g protein which very well meets the requirement of 350 g per day protein required for maintenance of adult cattle of ~500–600 kg weight (Table 9.2). Macro- and micronutrients are required for good health and immunity in cattle. Spent malt is a rich source of macro- and micronutrients – 2 kg spent malt provides nutrients at par or more than the recommended 100 g mineral mixture per day.

There is interesting story of popularization of spent malt as animal feed in internationally known watershed at Kothapally. Actually, ICRISAT and SABMiller India were into a Memorandum of Agreement (MoA) in August 2009 for collaborative watershed activities in Fasalvadi, Chakriyal, Venkatakishtapur and Shivampet villages which were later expanded to ten villages. In this regard, to learn the watershed intervention, Fasalvadi farmers had an exposure visit to Adarsha watershed at Kothapally, during which, farmers from Kothapally came to know about the spent malt initiative and its benefits realized by Fasalvadi women. Kothapally is a village with milk production activity of around 2100 l per day. In this context, lead women farmers in Kothapally watershed realized opportunities of improving milk production through getting spent malt from nearby SABMiller brewery. They requested ICRISAT to discuss with SABMiller India and establish a spent malt initiative. ICRISAT intervened with its CSR partner to facilitate and launch Spent Malt activity in Kothapally village on 17 June 2013. Women farmers were organized into SHGs to handle all logistics of transportation from the factory, and distribution amongst fellow farmers and a successful business model was implemented. Training component was handled by ICRISAT, and major points to take care in spent malt use are as under:

- Spent malt (wet) to be consumed within 24 h. Thereafter, it gets fermented and sour.
- Not to be fed to cattle after 48 h – worms may get developed and cattle health may be affected.
- Fresh spent malt needs to be dried for storage and use later on.
- Quantity to be fed is 4–5 kg spent malt day^{-1} $animal^{-1}$ (2–2.5 kg in the morning and same in the evening).

The basic requirements in this initiative are as follows:

- Vehicle arrangement for lifting spent malt from brewery to respective village
- Place with rooftop for unloading and storing spent malt
- Plastic drums (200 l size) for storing spent malt
- Buckets/baskets for unloading spent malt
- Weighing balance for distribution of spent malt to farmers
- Inventory books for maintaining disbursement details, etc.

Tejasri women's SHG in Adarsha watershed, Kothapally village, in Ranga Reddy district, is handling the spent malt-based activity which is a group of 12 women members. Around 96 households in the watershed purchase spent malt to feed around 559 milch animals (Fig. 9.3). Daily, around 2580 kg spent malt is used to feed cattle. With use of spent malt as animal feed, farmers have observed increased milk production of about 2 l per animal per day with improved fat content. Due to this, the gross income in the village is increased by about Rs. 46,000 per day (about Rs 36,000 net income) on account of increased milk production in the village. On a monthly basis, more than Rs 11,000/– net income is increased per household of participating farmers. Tejasri group that handles the activity procures spent malt at the rate of Rs 2.75 per kg and sells at the rate of Rs 4 per kg. Members use Rs 1.25 per kg for transportation and handling charges by the group. Through this, member handling day-to-day operations gets around Rs 10,000/– per month income and contributes Rs 1000/ – for the group corpus fund.

Fig. 9.3 Scaling-up of spent malt as animal feed through women SHG in Kothapally

With the success of model in Kothapally, it has captured the attention of many stakeholders and now scaled-out to other locations like Neemrana (Rajasthan), Murthal (Haryana), Hyderabad (Telangana), Mysore (Karnataka), etc.

9.3   Nursery Raising and Nutri-kitchen Gardens

Nursery raising of fruits, plantation, vegetable and ornamentals is a potential opportunity for women farmers as a livelihood activity. Women in Kothapally watershed adopted nursery raising of fruits and plantation crops as a livelihood activity. During the watershed program, women raised nurseries and supplied 2500 fruit trees and teak plants along with about 50,000 *Gliricidia* saplings planted on bunds for generating N-rich organic matter. Nurseries in horticulture plants is important area for income generation for women due to the large scope of horticulture sector in total per cent share of around 30% in agricultural output and a key area to achieve desired doubling of farmers' income and resilience in the drylands. In horticulture sector, per cent share of production of fruits and plantation crops is quite significant at 37%. Raising ornamental plants for city markets is also a big opportunity. In view of low soil organic carbon levels of farmers' fields and low quantities of recyclable organic carbon, biomass generation through nitrogen-rich green manure plants is also need of the hour. In this context, *Gliricidia* plantations on the farm boundaries have proved very beneficial for adding carbon and nutrients to the fields through chopping leaves before rainy, post-rainy and summer seasons. On-station watershed studies at ICRISAT have shown that *Gliricidia* loppings provide 30 kg N ha^{-1} year^{-1} without adversely affecting crop yield.

Kothapally watershed has pilot tested the model of nutri-kitchen gardens through which women cannot only improve nutrition of household but also earn income (or save expenditure) through sale of vegetables. Women are provided seeds of vegetable for cultivation in 10–20 m^2 as kitchen gardens along with know-how of cultivation. Most women use house-made compost for vegetable production. Nutri-kitchen garden kits with different vegetable crops (tomato, brinjal, okra, bottle gourd, bitter gourd, ridge gourd, *Palak (spinach)* and *Amaranthus*) were provided to 110 households every year (2016–2017 and 2017–2019) in Kothapally village to grow vegetables in their backyard for their household consumption resulting in saving expenditure on purchase of vegetables. These 110 households produced about 3000 kg of vegetables. The average household production is about 28 kg of vegetables with a saving of around Rs 800/family while improving household nutrition.

9.4 Composting: Recycling Wastes

With awareness about health, environment and resource use, the demand for organic-based products is increasing. Kothapally watershed is very close to Hyderabad city and better positioned to fulfil the huge demand from the city for compost use in ornamentals and kitchen or roof gardens in addition to use in farmers' agricultural fields. With such huge scope, women in Kothapally have adopted composting of residues and household waste as a remunerative business activity.

Composting is the technology for conversion of bulky organic wastes into low-volume nutrient-enriched and stable product. Traditional composting (farmers' practices of heaping straw and dung) is very time consuming and relatively less effective. In such a case, using half decomposed compost/manure/plant residue creates many plant nutrient and pest-related problems, rather than benefits. Vermicomposting is one of the tested technologies to effectively recycle on-farm wastes to produce quality compost for use in crop production (Chander et al. 2013, 2018; Wani et al. 2014a, b). Vermicomposting hastens the decomposition process through physical breakdown of the raw biomass coupled with mixing of vast spectrum of microbes with the biomass while passing through earthworm gut. The microbes of earthworm gut are highly potential in digesting the organic materials as well as polysaccharides (Aira et al. 2007; Zhang et al. 2000). Apparently high microbial activity under composting have an indirect role in improving compost nutrient quality by nitrogen fixers, nitrifiers and sulphur oxidizers (Richardson and Simpson 2011) and may also synthesize chemicals which act as plant growth hormones (Pizzeghello et al. 2001; Ghosh et al. 2003; Tomati et al. 1988).

In Kothapally watershed, the composting activity was initially adopted by 10 women farmers; however, by present day, 60 women farmers are involved in it (Fig. 9.4). One unit produces around 2500 kg compost in a year. Farmers get a price of about Rs 4/– per kg compost, and thus each person is able to earn around Rs 10,000/– a year through this activity. This side activity not only brings incomes to women farmers but also recycles household and on-farm wastes which otherwise do not find any effective alternate use except creating a nuisance. This activity also contributes to cleanliness drive in the village. One of the SHGs, namely Shivaganga group, is also engaged in making vermiwash through making outlets for collection of washings in composting unit. Per unit 150–200 l vermiwash is produced and is sold at Rs 4/– per litre. It is quite popular with vegetable farmers to improve quantity and quality of the produce (Wani et al. 2014b; Chander et al. 2013).

Fig. 9.4 Recycling of on-farm wastes into compost using vermicomposting technology in Kothapally

9.5 Improved Food, Nutrition and Livelihood Security

With the watershed interventions in Kothapally, food production increased as a result of diversification from dominant non-food crops to food crops like maize, sorghum and pigeon pea. Kothapally was predominantly a cotton-growing area prior to project implementation. The area under cotton was 200 ha in 1998. Maize, chickpea, sorghum, pigeon pea, vegetables and rice were grown in very limited area. After 4 years of activities in Adarsha watershed, the area under cotton cultivation decreased from 200 to 80 ha (60% decline) with simultaneous increases in maize and pigeon pea. The area under maize and pigeon pea increased more than threefold from 60–200 to 50–180 ha, respectively, within 4 years. The area under chickpea also increased twofold during the same period. With enhanced in situ and ex situ water availability, farmers started cultivating vegetables. Alongside, the productivity of crops increased by two- to threefold in crops like maize and sorghum as compared with the base year during 1998. All these changes brought in enhanced and diversified availability of food along with surpluses for income enhancement. With the changes scenario, migration of villagers stopped and brought in enhanced social security to family and women as such.

Moreover, the linkages of soil fertility management and fertilizer use with food quality and nutrition are well established (Chander et al. 2013; Wani and Chander 2016; Wani et al. 2017). In this regard, need-based fertilizer use has improved food nutritional quality and effectively reaching out to the children and women in the watershed. The impacts of soil health management are far beyond grain production and quality, especially in a predominant crop–livestock farming system in the watershed where crop residue serve as important feed components and their quantity and quality is positively affected with such a balanced fertilizer use strategy and HYVs in the watershed.

9.6   Other Allied Enterprises

Watershed experience highlighted clearly that household and women incomes as such can be significantly enhanced by shortening the value chain and strengthening networks for primary processing at farm level. In this regard, dal processing is a major activity adopted by women SHGs. Village seed bank is also an important activity where women are involved. This activity not only brings income to women but also ensures availability of good quality of seed for cultivation leading to higher production in the village. With support of ICRISAT, women SHGs also undertook specialized activities like *Helicoverpa* nuclear polyhedrosis virus (HNPV) production for minimizing pest damage in crops like cotton, pigeon pea and chickpea. Alongside, the project has given high priority to training village-level scouts to identify various pests and their natural enemies in different crops. With the economy picking up in the village, women have adopted other non-farm activities as well like petty shops as livelihood source for them.

9.7   Collectivization and Market Linkages

Small business size and little bargaining power of women farmers in the country is the major cause for most of the problems. Hence, collectivization of women producers is one of the most effective pathways to address the many challenges, and hence women in the watershed are organized into self-help groups (SHGs). The SHGs enabled reaping the benefits of economies of scale, reduce the transaction costs, improve profit margins and effectively manage risks and uncertainties. A strong stewardship for capacity building and strengthening knowledge base of women farmers helped in formulating good business plans and management. Organizing as SHG has not only enabled ease of business doing but also facilitated market linkages in a business model. With large number of women organized into milk production, Reliance has opened a milk collection centre which provides competitive price to the women. Similarly, other women are organized into SHGs around activities like composting, dal processing, seed banks, nursery raising and small-scale vegetable production.

9.8   Awareness and Capacity Building

Amongst others, one of the reasons for women lagging behind is the knowledge gap between 'What to do' and 'How to do it'. In view of human resource constraints and poor knowledge delivery system, i.e. extension system, the challenge to reach out to all women farmers is a huge one. The Kothapally watershed demonstrated the model that the information delivery mechanism can be strengthened by utilizing the services of practicing women farmers in the villages as lead farmers or farmer facilitators who stay in the villages and can effectively transmit knowledge and bring the fellow farmers on the board. The lead farmers across SHGs were given exposure visits and thorough trainings for various livelihood options, and they are those who transmitted it to the fellow farmers in SHGs. Exposure visits to ICRISAT campus and breweries and hands-on training courses were part of capacity building programs along with day-to-day hand-holding support during the watershed program implementation period during 1999–2003. This participatory program has developed women leaders with desired capacity to take forward various programs and has managed well even after completion of the watershed project in 2003. Women farmers have not only sustained the livelihoods but evolved over time and expanded the enterprises by seeking help of experts as and when needed. With progress over the years, Kothapally has become a bright spot for exposure visits for women from other regions to learn from and engage in livelihood activities to increase their incomes.

With increasing connectivity through Digital India initiative of the Government of India, there is wide scope for decision-making, monitoring, impact analysis and knowledge dissemination using ICT. As trained human resource is a major constraint, various ICTs are available which can bridge the gap between women farmer and knowledge generator. Rapidly evolving information technology industry and favourable environment for ICT in agriculture are giving a great boost to agricultural extension.

9.9   Summary and Key Findings

It has been demonstrated in a study that mere engagement of women in watershed activities does not benefit women unless the income-generating activities are brought in to benefit the women. It was noted that women having more income in the watersheds at their disposal were having more confidence, self-esteem and decision-making authority in the family (Sreedevi et al. 2004; Sreedevi and Wani 2007). Initially, it was noted that Kothapally watershed was at the lowest rung on the ladder amongst the three watersheds studied for benefits to women. Taking these results in to consideration, subsequently, more income-generating activities were promoted in the Kothapally watershed.

Successful models of mainstreaming women farmers and increasing their incomes have been put in place in Kothapally watershed through women-focused interventions like animal rearing, spent malt as animal feed,, kitchen gardens, composting, value addition and non-farm-based livelihoods. With direct benefits to women and family as such, these need to be scaled out in other geographies. Milk production is in general a big activity in the domain of women, and strategic marketing interventions like cooperatives in India have linked women to the markets to some extent, but a lot more need to be done for coverage across the country. Private players as done in Kothapally also need to be roped for market linkages in many areas. The next opportunities lie in increasing production as is demonstrated in Kothapally though addressing the issues of fodder scarcity in drylands, making available the concentrates, breed improvement and expansion. In current times, with focus to double farmers' incomes, primary processing at farm level is suggested to retain the maximum value share with farmers. Kothapally watershed has piloted dal processing as women-focused enterprise, and there are similar many other opportunities where women farmers can be roped in for value addition. Poor financial condition and poor risk taking ability are major deterrents for majority of women farmers for the infrastructural and marketing requirements. To address the issues of family nutrition and income, promotion of kitchen gardens in rural areas could be a very important activity as is demonstrated in Kothapally. With the economy picking up with various interventions in the watersheds, there are other non-farm activities generated where women need to be roped in. In most of the interventions, a favourable policy to support financially and address risks through collectivization and market linkages is need of the hour. A framework of capacity building and hand-holding support is required in the policy to take forward the cause of mainstreaming women farmers and improving their incomes.

Acknowledgement We acknowledge the SHGs and farmers in general in the watershed in providing us with the updated data sets. The financial resources provided by the erstwhile undivided Andhra Pradesh Government, the Asian Development Bank, are gratefully acknowledged.

References

Aira, M., Monroy, F., & Dominguez, J. (2007). Earthworms strongly modify microbial biomass and activity triggering enzymatic activities during vermicomposting independently of the application rates of pig slurry. *Science of the Total Environment, 385*, 252–261.

Blümmel, M., Samad, M., Singh, O. P., & Amede, T. (2009). Opportunities and limitations of food–feed crops for livestock feeding and implications for livestock–water productivity. *Rangeland Journal, 31*, 207–213.

Chander, G., Wani, S. P., Sahrawat, K. L., Kamdi, P. J., Pal, C. K., Pal, D. K., & Mathur, T. P. (2013). Balanced and integrated nutrient management for enhanced and economic food production: Case study from rainfed semi-arid tropics in India. *Archives of Agronomy and Soil Science, 59*(12), 1643–1658.

Chander, G., Wani, S. P., Gopalakrishnan, S., Mahapatra, A., Chaudhury, S., Pawar, C. S., Kaushal, M., & Rao, A. V. R. K. R. (2018). Microbial consortium culture and vermicomposting technologies for recycling on-farm wastes and food production. *International Journal of Recycling of Organic Waste in Agriculture*. https://doi.org/10.1007/s40093-018-0195-9.

Ghosh, S., Penterman, J. N., Little, R. D., Chavez, R., & Glick, B. R. (2003). Three newly isolated plant growth promoting bacilli facilitate the seedling growth of canola, *Brassica campestris*. *Plant Physiology and Biochemistry, 41*, 277–281.

Haileslassie, A., Blümmel, M., Clement, F., Descheemaeker, K., Amede, T., Samireddypalle, A., Acharya, N. S., Radha, A. V., Ishaq, S., Samad, M., et al. (2011). Assessment of livestock feed and water nexus across mixed crop livestock system's intensification gradient: An example from the Indo-Ganga Basin. *Experimental Agriculture, 47*, 113–132.

Haileslassie, A., Blümmel, M., Wani, S. P., Sahrawat, K. L., Pardhasaradhi, G., & Samireddypalle, A. (2013). Extractable soil nutrient effects on feed quality traits of crop residues in the semiarid rainfed mixed crop–livestock farming systems of Southern India. *Environment, Development and Sustainability, 15*, 723–741.

Pizzeghello, D., Nicolini, G., & Nardi, S. (2001). Hormone-like activity of humic substances in Fagus sylvaticae forest. *The New Phytologist, 151*, 647–657.

Richardson, A. E., & Simpson, R. J. (2011). Soil microorganisms mediating phosphorus availability. *Plant Physiology, 156*, 989–996.

Shiferaw, B., Anupama, G. V., Nageswara Rao, G. D., & Wani, S. P. (2002). *Socioeconomic characterization and analysis of resource-use patterns in community watersheds in semi-arid India* (Working Paper Series No. 12). Patancheru: International Crops Research Institute for the Semi-Arid Tropics. 44 pp.

Sreedevi, T. K. and Wani, S. P. (2007). Leveraging institutions for enhanced collective action in community watersheds through harnessing gender power for sustainable development. Paper published in book titled" Empowering the poor in the era of knowledge economy. (ed. Srinivas Mudrakartha, VIKSAT, Ahmedabad). pp.27–39.

Sreedevi, T. K., Shiefaw, B., & Wani, S. P. (2004). *Adarsha watershed in Kothapally: Understanding the drivers of higher impact* (Global theme on agroecosystems report no. 10). Patancheru: International Crops Research Institute for the Semi-Arid Tropic. 24 pp.

Tomati, U., Grappelli, A., & Galli, E. (1988). The hormone-like effect of earthworm casts on plant growth. *Biology and Fertility of Soils, 5*(4), 288–294.

Wani, S. P., & Chander, G. (2016). Role of micro and secondary nutrients in achieving food and nutritional security. *Advances in Plants & Agriculture Research, 4*(2), 131.

Wani, S. P., & Shiferaw, B. (2005). *Baseline characterization of benchmark watersheds in India, Thailand and Vietnam* (Global theme on agroecosystems report no. 13). Patancheru: International Crops Research Institute for the Semi-Arid Tropics. 104 pp.

Wani, S. P., Dixin, Y., Li, Z., Dar, W. D., & Chander, G. (2012). Enhancing agricultural productivity and rural incomes through sustainable use of natural resources in the semi-arid tropics. *Journal of the Science of Food and Agriculture, 92*(2012), 1054–1063.

Wani, S. P., Chander, G., & Sahrawat, K. L. (2014a). Science-led interventions in integrated watersheds to improve smallholders' livelihoods. *NJAS – Wageningen Journal of Life Sciences, 70-71*, 71–77.

Wani, S. P., Chander, G., & Vineela, C. (2014b). Vermicomposting: Recycling wastes into valuable manure for sustained crop intensification in the semi-arid tropics. In R. Chandra & K. P. Raverkar (Eds.), *Bioresources for sustainable plant nutrient management* (pp. 123–151). Delhi: Satish Serial Publishing House. Available: https://www.researchgate.net/publication/270528480_Vermicomposting_Recycling_Wastes_into_Valuable_Manure_for_Sustained_Crop_Intensification_in_the_Semi-Arid_Tropics ; http://hdl.handle.net/10568/75837.

Wani, S. P., Chander, G. and Anantha, K. H. (2017). Enhancing resource use efficiency through soil management for improving livelihoods. In: Adaptive soil management: From theory to practices, Chapter: 19 (Rakshit, A., Abhilash, P. C., Singh, H. B., Ghosh, S. Ed.). Springer Singapore. pp 413–451.

Zhang, B. G., Li, G. T., Shen, T. S., Wang, J. K., & Sun, Z. (2000). Changes in microbial biomass C, N, and P and enzyme activities in soil incubated with the earthworms *Metaphire guillelmi* or *Eisenia fetida*. *Soil Biology and Biochemistry, 32*, 2055–2062.

Chapter 10
Increasing Incomes and Building Climate Resilience of Communities Through Watershed Development in Rainfed Areas

K. H. Anantha, S. P. Wani, and D. Moses Shyam

Abstract This chapter documents the 28% increase in family incomes of watershed over non-watershed villages due to the integrated watershed development livelihood model. Kothapally is a unique peri-urban village in the vicinity of Hyderabad with 400 households mainly cultivators. Further, it has also demonstrated that through watershed development resilience of the communities during drought years was built and no migration took place as share of crop income remained constant, whereas in non-watershed villages it dropped by 75% and people including the farming households had to migrate for their livelihoods, and proportion of income from non-agriculture activities increased dramatically to 74%. Suitability of the integrated watershed livelihood approach strongly indicated the approach to be scaled-up not only for increasing production and income but also for contributing substantially to the SDGs, viz., zero hunger, reducing poverty, climate change interventions, and women empowerment.

Keywords Watershed management · Increased rural incomes · Income diversification · Rainfed areas

10.1 Introduction

Watershed development programs in India are aimed at improving and sustaining productivity and production potential of the dry and semi-arid regions of the country at higher levels, through the adoption of appropriate production and conservation techniques. In recognition of rural distress especially in rainfed areas,

K. H. Anantha (✉) · D. Moses Shyam
ICRISAT Development Centre, Research Program Asia, International Crops Research Institute for the Semi-Arid Tropics (ICRISAT), Hyderabad, Telangana, India
e-mail: k.anantha@cgiar.org

S. P. Wani
Former Director, Research Program Asia and ICRISAT Development Centre, International Crops Research Institute for the Semi-Arid Tropics (ICRISAT), Hyderabad, Telangana, India

© Springer Nature Switzerland AG 2020
S. P. Wani, K. V. Raju (eds.), *Community and Climate Resilience in the Semi-Arid Tropics*, https://doi.org/10.1007/978-3-030-29918-7_10

government, donors, and development partners have devoted substantial resources to develop and promote rainfed areas for sustainable intensification of agriculture and rural livelihoods through watershed development programs. The aim is also to meet the needs of rural communities for food, fuel, fodder, and timber and thereby reduce pressure on the natural environment (GoI 2008; Wani et al. 2008). In view of their potential for growth and for improving income levels and the natural resource base of the disadvantaged regions of the country, watershed development programs have been accorded priority in India's development plans and by a number of donor agencies (GoI 2007; Joshi et al. 2009). The impacts of watershed programs in India revealed that watershed projects yielded multiple exemplary benefits in terms of economic, sustainability, and equity parameters (Joshi et al. 2008) and reconfirm that watershed projects are economically viable and generate substantial economic, social, and environmental benefits and justify the investment in watershed programs as income levels were raised within the target domains (Palanisami et al. 2009; Wani et al. 2011, 2012).

Watershed development programs assume greater significance in India where 56% of the cropped area is rainfed and is characterized by low productivity, water scarcity, degraded natural resources, and widespread poverty. These initiatives have produced notable impacts in terms of increasing income and environmental protection (Wani et al. 2011, 2012; Garg et al. 2012). Further, it would guarantee more food, fodder, fuel, and livelihood security for those who are at the bottom of the rural income scale (Palanisami et al. 2009; Wani et al. 2012). The present scenario thus clearly points to the need for adoption of science-led interventions leading to efficient and sustainable use of natural resources to improve agricultural productivity and rural livelihoods to alleviate poverty in semi-arid regions.

In this backdrop, the present chapter discusses the opportunities to enhance the household income in the watershed context of Indian semi-arid region using data collected during the years 2010 and 2018. The chapter specifically focuses on the patterns of income diversification and their determinants in the context of semi-arid watershed and suggests policy recommendations for future consideration.

10.2 Watershed Management and Rural Livelihoods

Watershed management in India has undergone several structural changes to address and include emerging issues of hunger and poverty (Wani et al. 2006; Raju et al. 2008; GoI 2008). Earlier experiences from the various watershed projects have indicated that a straightjacket approach did not yield desired results and mix-up of individual and community-based interventions is essential (Joshi et al. 2005; GoI 2008). Multidisciplinary teams are involved to provide all the technical expertise to solve the problems at the community level. The benefits are transparent and distributed equally well among the community members including women. As a result, the level of participation has improved. This approach ensured participation and the watershed is considered as an entry point for improving the livelihoods of the people (Wani et al. 2008).

Reviews of watershed experiences in the 1970s and 1980s identified the lack of attention to farmers' objectives and farmers' knowledge as the important reason for these failures. In contrast, where user participation was incorporated, performance of the watershed projects improved (Kerr et al. 2000). As a result of these lessons, many participatory watershed development interventions were designed and implemented with explicit involvement of users and sought to address their livelihood concerns. The watershed development program is now planned, implemented, monitored, and maintained by the watershed communities. To bring about uniformity in programs being implemented by various agencies in India, the WARASA-Jan Sahbhagita Guidelines were formulated in conformity with the "Common Approach/ Principles for Watershed Development" agreed upon by the Ministries of Agriculture and Rural Development, Government of India. The National Watershed Development Project in Rainfed Areas (NWDPRA) was considerably restructured during the ninth Five-Year Plan with greater decentralization and community participation, higher degree of flexibility in choice of technology, and suitable institutional arrangements for ensuring long-term sustainability. The 1994 guidelines provided special emphasis to improve the economic and social conditions of the resource-poor and the disadvantaged sections of the watershed community. While few rigorous evaluations of this experience exist, case studies suggest that their performance has been better, at least in terms of governance and technology adoption (Sreedevi et al. 2006, 2008; Pathak et al. 2007; Wani et al. 2008). Focusing watershed interventions more directly on the needs of local communities is likely to make their outcomes more pro-poor. Recently, watershed management programs sought to embed the livelihood approach and local participatory planning processes initiated as part of the participatory watershed initiatives within broader social and political processes more explicitly (FAO 2006; GoI 2008). Special attention is placed on strengthening and supporting the poor in their ability to participate in project planning and implementation process and to diversify their livelihood strategies using new science tools (GoI 2008). While motivation for diversifying livelihood strategies may be either positive or negative, a growing number of studies suggest that such strategies do have beneficial effects on rural livelihoods (Sreedevi et al. 2006, 2008; Palanisami et al. 2009). Therefore, the impacts of integrated watershed management programs may have significant implications for the welfare of the poor.

10.3   Data and Methodology

10.3.1   Study Area

Kothapally village is located in Ranga Reddy district of erstwhile Andhra Pradesh state in Southern India. Ranga Reddy district, like many other parts of Andhra Pradesh, has seen a dramatic change in overall development in the past decade. Kothapally is inhabited by nearly 400 (270 HHs at the start of the watershed – 1999) farming households, either as owner-cultivators or landless laborers. It covers

465 ha, of which 430 ha are cultivable and 35 ha wasteland. The annual average rainfall in the area is about 800 mm (85% of it occurs from Jun to Oct). The watershed is characterized by undulating topography (the slope of the land is about 3%) and predominantly black soils which range from shallow to medium deep black with a depth of 30–90 cm. Farmers diversify their cropping pattern across a number of crops grown during two seasons: rainy and post-rainy. The crops grown include sorghum, pigeon pea, black gram (*Phaseolus mungo*), maize (*Zea mays*), paddy (*Oryza sativa*), cotton (*Gossypium hirsutum*), sunflower (*Helianthus annuus*), and vegetable bean (*Dolichos lablab*), mostly under rainfed conditions. Paddy, sorghum, sunflower, and vegetables are grown in the post-rainy season using residual moisture and supplementary irrigation. There is also some area for growing turmeric (*Curcuma longa*), onion, and paddy which uses tube well irrigation. Recently, ICRISAT introduced chickpea (new varieties) grown in the post-rainy season, and the area for growing maize has substantially increased, often at the cost of cotton. The interventions were implemented through a multi-institutional consortium and the local community. There is also a substantial livestock population that is equally dependent on safe water supply.

10.3.2   Baseline Characterization[1]

The baseline survey in Adarsha watershed was carried out in April 1999 (Shiferaw et al. 2002). Based on total land ownership, the 270 farmers in the watershed were stratified into three groups: small (less than 1 ha, 136 households), medium (1–2 ha, 60 households), and large (above2 ha, 74 households) farmers. A certain number of households were randomly selected from each group to arrive at a sample size of 54 households. The data shows that 22% of the surveyed households were women farmers. Family size exhibited a wide variation ranging between 2 and 25 persons. The average family size was 7.33 persons. About two-thirds of the households had a family size less than the average. The remaining one-third had household sizes above the average. About half of the households also had a family size of less than five. The average number of males was 3.74 and females 3.59. The average weighted labor force per household was 4.32 persons, indicating a worker-consumer ratio of about 60%.

The dependency ratio, i.e., the number of non-working members per working family member, was 0.78, indicating a high degree of dependency. This implies that every working member of the family supports on an average 0.78 dependents, which include children and senior citizens. The caste composition of the surveyed households was as follows: Backward Caste (54%), Scheduled Caste (20%), Muslim (12%), and others (14%). The average age and level of education of the household head were about 45 and 2.63 years, respectively. About 70% of the household heads were uneducated. The average level of education in the family was slightly higher

---

[1] This section is based on Shiferaw et al. (2002).

at 3.13 years. The number of illiterate family members in every household averaged 2.95, indicating 40% illiteracy within the household (Shiferaw et al. 2002).

Although, Kothapally had a good road network, the average distance to the nearest market-town (20 km away) implies major transportation costs for farmers. Agriculture was stated as the main source of income for all households, indicating the dire lack of other income-earning opportunities in the area. The respondents were all cultivators, with an average land ownership of 1.43 ha per household, translating into a land-person ratio of about 0.195 ha. This is a very negligible size of land and requires serious intensification and multiple cropping to provide the required food security to the household. About 80% of the farmland was non-irrigated. The total owned cultivated land area was 1.295 ha, distributed into dry land (1.012 ha, 78%) and irrigable land (0.283 ha, 22%). All the respondents indicated that the soil type was black, with a depth ranging from 0.5 to 3.5 m (average depth of 2.19 m). This indicates that the farmers' estimates and perceptions of soil depth in the area were quite high. Farmers' responses showed that more than 90% of the farms had a soil depth of more than 1 m and more than half of them had a soil depth of more than 2 m.

In terms of access to irrigation, about 60% of the farmers revealed they had no source of irrigation. Farmers used different types of water-harvesting methods. A third of the respondents used tube wells, while the rest used open wells and tanks as sources of irrigation. As community efforts toward investments in check dams and other structures to retain runoff water succeed, there are reports that the groundwater table, as well as the water level, in private wells is rising. There is an interesting contradiction between community ownership of water conservation investments (e.g., check dams) and private tapping of groundwater by drilling wells near check dams. If unregulated, this may increase the exploitation of groundwater and has the potential to undo community benefits and enhanced ecological services of watershed investments.

Seasonal land use patterns indicate that the average operated area in the rainy season was 1.295 ha, while the average for the post-rainy season was 0.275 ha. Very little of the operated area was under temporary or permanent fallow, indicating a high intensity of land use and rotation value for farmland. For some inexplicable reason based on the available data, it was found that none of the households participated in local land rental markets through fixed-rental leasing in/out or through share cropping. All the households were self-sufficient in land use, which may be due to a sampling bias or serious imperfections in village land markets. The small landholdings seemed to leave little in terms of surplus to rent out to other households. This seems so from the distribution of land, which indicates that only 16% of the respondents had landholdings above 2 ha and about 55% had below 1 ha. The remaining 29% had landholdings between 1 and 2 ha.

A majority of the farmers in the area are mixed crop-livestock producers. The major types of livestock included cattle, buffaloes, goats, sheep, and poultry. About 72% of the respondents owned some livestock in addition to indulging in crop-production activities. About 48% of the households also owned bullocks (including improved and local breeds). About 37% of the households owned a pair of bullocks needed for transportation and cultivation. About 6% owned more than a pair of bullocks, while about the same percentage owned only one bullock. Very few households (11%) owned any milching cow, but about 35% owned she-buffaloes. The average ownership of different types of animals was 1.05 (bullocks), 0.11 (milching cow), 0.13 (young cattle), 0.5(she-buffaloes), 0.43 (young buffaloes), 0.76 (goats), 0.83 (sheep), and 0.65 (poultry). After bullocks that are needed for transporting goods and cultivation, goats are the most popular small stock kept on the farm. About 41% of the households engage in raising goats. Only a few households (4%) raise sheep on the farm.

Apart from livestock, farmers also possessed other assets and implements (such as tractors, bicycles, plows, seed drills, and bullock carts) mainly used in crop and livestock production. The average farm equipment and related wealth of the sample households was Rs. 15374, of which 57% possessed assets worth less than Rs. 10,000. Some 35% owned assets worth between Rs. 10,000 and 25,000. In terms of important assets, nearly 98% of the households did not own any tractors. Hence the average tractor ownership was only 0.0185. On the other hand, more than 68% of the households owned a seed drill and 88% owned a sprayer.

The baseline survey included questions regarding crop production, cropping patterns, and input and output relationships. Data on cropping pattern indicates that some crops were grown as sole crops, while others were grown as intercrops. In Adarsha watershed, crops grown as monocultures included cotton, paddy, vegetable bean, maize, sorghum, sunflower, and turmeric. Other crops mainly grown as intercrops on the same field included sorghum, black gram, and pigeon pea.

In Adarsha watershed, the most commonly grown crops during the rainy season were intercrops consisting of sorghum, black gram, and pigeon pea. About 60% of the surveyed farmers in the area reported growing these crops as intercrops. Based on farmers' responses, the average share of land allocated to the different crops in the intercropping system was 80% for sorghum and 10% each for black gram and pigeon pea. This shows that pulses actually occupy a small proportion of the land, which is mainly allocated to a cereal (sorghum). The results also show that relatively fewer households grew other crops (as single stands): cotton (30%), paddy (33%), vegetable bean (31%), sunflower (6%), tomato (7%), and turmeric (13%).

The average area cultivated to each of these crops reveals that the largest share of cropland was allocated to the sorghum-pigeon pea intercrop. Only a small proportion (less than 20%) of the land area under these crops was irrigated. The average level of fertilizer use was unreliable as data was perhaps missing (due to empty fields) for a number of sample farmers. The average provided was based on recorded positive levels of use and on the assumption that empty fields meant non-use. The results were on the higher side, indicating the substantial use of DAP and urea fertilizers per ha for all crops, except tomatoes. The average yield of sorghum was about 1100 kg ha^{-1}, black gram 110 kg ha^{-1}, and pigeon pea 203 kg ha^{-1} in the intercropping system. The average yield of paddy grown mainly with supplementary irrigation during the rainy season was about 5486 kg ha^{-1}, while that grown during the post-rainy season was about 4480 kg ha^{-1}. Vegetable bean, another important crop grown by a third of the farmers, gave average yields of about 2890 kg ha^{-1}. The share of farmers growing cotton was of a similar order of magnitude as that of paddy and vegetable bean, and the average yields were about 1800 kg ha^{-1}.

The variability and stability of yields were measured by the coefficient of variation (CV), the standard deviation as a percentage of mean yield. There was considerable variation in yield levels attained among different farmers, perhaps reflecting the effects of land quality and input intensities in growing these crops. The baseline data lacks such detail at the plot level and the relative contribution of variable and fixed factors in determining crop production in the area cannot be accounted for. One major factor associated with variability in crop yields is the level of irrigation used. The variability was greater in the case of non-irrigated crops and tended to decline with the share of land irrigated. Hence, the variability in yields was highest in vegetable bean and cotton as well as in the sorghum-pigeon pea intercrops grown under rainfed conditions. The variability in yield was less in paddy grown with irrigation.

A look at the average yields will reveal that wheat and chickpea grown as sole crops had higher yields than the intercrops. As sole stands, post-rainy-season wheat yields were about 1200 kg ha^{-1}, while chickpea yields were about 930 kg ha^{-1}. In the rainy season, the average soybean yield from farmers' fields was about 760 kg ha^{-1}, whereas paddy provided about 600 kg ha^{-1}. The results seem to show a relatively lower variability in yields among farmers in this area than in Adarsha watershed, perhaps due to the higher and more reliable rainfall pattern and better soils. For the most commonly grown crops, variability in grain yield seemed to be higher for crops grown in the post-rainy season, perhaps indicating the importance of access to supplementary irrigation. Better data is needed to estimate the relative profitability of crops and cropping patterns, partial effects of improved input usage, the quality of soil, and the effect of irrigation on crop yields and variability of income.

The poor quality of household expenditure on factors of production and the lack of records on farmers' incomes from sources other than cropping have now made it difficult to estimate farmers' net incomes. In fact, the survey even failed to ask farmers about their earnings from livestock production; only livestock wealth at the beginning of the year was recorded. Often, computing gross returns from livestock requires data on changes in the stock of animals during a given year. Income from local farm and non-farm employment, petty trade, migration (remittances), etc. was not compiled. As water availability in the watershed increases and land productivity goes up, the level of production risk faced by households may change. This may create new crop-livestock production patterns and increase possibilities for local employment. In view of the potential of watershed investments to create such employment and income-earning opportunities for landless households and small holder farmers in the area, data on income from livestock and non-farm sources need to be collected as part of future surveys. This implies that future surveys should not only increase the sample size but also include landless households. Monitoring changes in household income and livelihoods would be difficult without such a complete dataset.

10.3.3 Household Survey

This study is based on household data collected in 2010 from 120 farm households representing small and medium farmers in the watershed and non-watershed villages. For better comparison of the impacts of watershed intervention on rural livelihood diversification, five adjoining villages that did not benefit from the project were included in the panel survey to address attribution problems. The study also utilized data collected in 2018 through household census in the watershed. The farm households in the watershed and non-watershed villages are similar in both agroclimatic and socioeconomic conditions, allowing us to separate the likely impact of the interventions after controlling for village level and other fixed effects. The sample households were selected based on stratified sampling technique. A pre-tested standard structured household questionnaire was used to collect information on household composition, socioeconomic characteristics, consumption, income, and asset position including participation in different farm and off-farm activities. For the purpose of analysis, we disaggregated income sources into five major categories, viz., farm, livestock, casual village labors, business and service, and capital earnings.

Selected household characteristics are presented in Table 10.1. The average household size in watershed and control villages is less than five adult equivalents which are below the national average of five adult equivalents (GoI 2011). In the overall sample, 10% of households are women-headed; if we present disaggregated figures of watershed and non-watershed samples, the proportion is 5.1 and 15.3%, respectively. Average farm size in the watershed is 2.5 ha which is slightly higher than the control villages and consistent with the national average. Considerable livestock population is present in both watershed and control villages ranging from 1.5 total livestock unit (TLU) per household in watershed to 1.6 TLU in control villages. The major determinants of income diversification are average number of income sources and number of crops grown which are slightly higher in watershed villages compared to control villages.

Table 10.1 Profile of sample households in watershed and non-watershed villages

Particulars	Watershed	Non-watershed
Household size (adult equivalent)	4.7	4.5
Age of household head (years)	53	50
Education of household head (years)	3.5	2.3
Farm size (ha)	2.5	2.1
Livestock population (TLU)	1.5	1.6
Average no. of income sources (NIS)	3.4	2.8
Average no. of crops grown	2.6	2.4

Source: Intensive household survey, 2010

Table 10.2 Endowment of human capital (%) for economically active population

Level of education	Watershed villages			Non-watershed villages		
	Male	Female	Total	Male	Female	Total
No formal education	27.6	56.7	41.7	31.4	55.6	42.8
Attended primary school	22.0	16.7	19.4	19.8	17.6	18.8
Attended secondary school	40.2	24.2	32.4	30.6	23.1	27.1
Have college-level education	10.2	2.5	6.5	18.2	3.7	11.4
Total	100	100	100	100	100	100

Source: Intensive household survey, 2010

The endowment of human capital in the study area was estimated for economically active population (Table 10.2). It revealed that almost same level of human capital endowment exists in both the areas. However, difference exists in terms of population having attended secondary school and college-level education. In terms of secondary school, watershed villages have an advantage compared to non-watershed villages as nearly 32% of economically active population had attended secondary school, whereas in non-watershed villages this proportion is merely 27%. In terms of college-level education, non-watershed areas have comparative advantage over watershed villages. In terms of gender, more than 50% of women have no formal education. However, the situation is comparatively better off in watershed villages.

10.3.4 Methodology

Two different methodologies are available in the development economic literature to analyze pattern of income diversification. The first one is an income-based approach which is based on household income accumulated from different sources and the second one is asset-based approach (Babatunde and Qaim 2009). In the income-based approach, we concentrate on three different income-based measures, viz., number of income sources, share of off-farm income, and Herfindahl diversification index (HDI). The number of income sources is based on the participation of household members in different activities. The share of off-farm income concentrates on off-farm income. The HDI is a measure of overall diversification that takes into account not only the number of income sources but also the magnitude of income derived from them. The HDI is based on the Herfindahl index which originates from the industrial literature where it is used to measure the degree of concentration. It can also be used to measure the degree of concentration of income from various sources at the individual household level. It is then calculated as the sums of squares of income shares from each income source. The Herfindahl index as such is increasing in concentration, whereby households with perfect specialization have a value of one. Since our interest is in diversification, which is the reverse of diversification, we use the HDI, which is defined as one minus the Herfindahl index (Babatunde and Qaim 2009). Thus, households with most diversified income sources have the largest HDI and vice versa (Barnett and Reardon 2000). This methodology has been used in Minot et al. (2006), Ersado (2005), and Babatande and Qaim (2009) for analyzing the patterns of income diversification in the context of Vietnam and Nigeria, respectively. Hence, this study is complementary to the above studies in the context of watershed management in India.

10.4 Results and Discussions

10.4.1 Pattern of Income Diversification

In rural areas, households do not restrict themselves to any single income source for improving their livelihoods (Reardon and Vosti 1995; NABARD 2018). Instead, they depend on many sources to supplement the major source of income. In the study area, farm households depend mainly on agriculture. However, farm income is supplemented by many other sources such as wage income, petty business, service, and others, viz., remittances and pension. In this backdrop, it is critical to understand whether the watershed intervention is helpful in diversifying income sources to build social resilience of farm households. Further, the household participation rate in different income activities would provide insights into the factors determining income diversification. Therefore, it is necessary to understand the pattern of income diversification with attention to household participation in different income activities.

10.4.2 Household Participation in Different Activities

As discussed above, farm households depend on several activities to supplement their farm income. Therefore, it is essential to understand their participation in different activities. To reflect household living standards appropriately, the income quartiles are formed based on total household income. The definition of participation used in this paper is the receipt of any income by any household member from a particular activity. Table 10.3 depicts the rate of household participation in different income activities. It is clearly demonstrated from the data that farming emerges as the single largest income source where all households are involved both in watershed and non-watershed areas. However, off-farm income sources are also receiving greater attention more in non-watershed villages compared to watershed villages. Possible reasons for this kind of transformation in non-watershed areas are free

Table 10.3 Household (HH) participation (%) in different income activities by income quartile

Income sources	Watershed villages						Non-watershed villages					
	First N = 12	Second N = 14	Third N = 16	Fourth N = 17	All HHs N = 59		First N = 17	Second N = 16	Third N = 14	Fourth N = 12	All HHs N = 59
Farming	92	100	94	100	97		82	88	86	100	88
Livestock	8	0	25	35	19		6	6	43	50	24
Wage income	50	57	94	35	59		82	75	79	58	75
Business and service	8	14	13	71	29		6	19	43	58	29
Capital earnings	58	36	44	41	44		35	31	14	33	29

Source: Intensive household survey, 2010

Table 10.4 Contribution from different sources in watershed and non-watershed villages by income quartile

Income sources	Watershed village					Non-watershed villages				
	First	Second	Third	Fourth	All HHs	First	Second	Third	Fourth	All HHs
Farming	93.7	100.0	90.2	97.5	96.3	94.2	97.3	75.9	77.9	81.7
Livestock	6.3	0.0	9.8	2.5	3.7	5.8	2.7	24.1	22.1	18.3
Total farm income	65.3	67.1	46.3	61.6	59.0	27.5	28.0	30.5	33.2	31.2
Wage labor	78.2	76.3	83.6	13.9	45.3	79.3	77.3	59.3	30.9	50.3
Business and service	7.9	17.2	13.8	84.5	51.7	10.9	19.9	39.8	60.5	43.8
Capital earnings (remittances, pension)	13.9	6.5	2.6	1.6	3.1	9.8	2.8	0.9	8.7	5.9
Total off-farm income	34.7	32.9	53.7	38.4	41.0	72.5	72.0	69.5	66.8	68.8

Source: Intensive household survey, 2010

labor mobilization and less opportunities to be involved in the farming sector. Earlier studies observed that transformation of agriculture development in watershed villages from one season to three seasons attracted unskilled labor force from neighboring villages (Wani et al. 2003). This kind of spillover effect is possible due to watershed intervention.

Table 10.4 shows that all households derived income from farming activities which contributed 59% in watershed villages and about 31% in non-watershed villages. It is important to note that about 48% of households at all India level and 47% of households in Telangana are farming households, whereas in Kothapally, about 98.5% of households are farming households and only 1.5% households are non-farming households. Farming or crop production, by far subsistent in nature, is the single major source of income in watershed villages as it contributed to about 96% to total farm income. Off-farm income activities played a crucial role in non-watershed villages as these activities contributed nearly 69% to total household income. It is important to note that both poor and rich households alike derived equally from the agriculture sector in watershed villages, while poorer households derived higher share from off-farm households in non-watershed villages (Table 10.4). This reflected the equal distribution of resource endowments among small and large farmers in the watershed compared to control non-watershed villages. The erstwhile watershed activities in the village benefitted in terms of water availability and soil fertility through several soil and water conservation activities. As a result, farmers who concentrated on farming activities fare better in terms of net returns as compared to the control villages. At the same time, focus also was on non-farm activities to supplement farm income through undertaking wage work and petty business.

Table 10.5 Average household income (US$) by income quartile

Sources of income	First	Second	Third	Fourth
Watershed villages				
Farming	388	748	767	2364
Livestock	288	0	312	174
Farm income	380	748	671	1793
Wage income	316	489	825	594
Business and service	192	440	1020	1810
Capital earnings	48	66	56	58
Off-farm income	173	342	617	1028
Non-watershed villages				
Farming	168	333	458	1112
Livestock	144	128	291	631
Farm income	166	319	402	952
Wage income	374	791	892	1523
Business and service	720	817	1096	2983
Capital earnings	107	68	72	750
Off-farm income	314	614	870	1919

Source: Intensive household survey, 2010

However, there exists a large gap in terms of average household income between watershed and non-watershed villages. Inter- and intra-village differences also existed in terms of average household income as poor households received less income than rich households. However, farm income contributed a major share in watershed villages compared to non-watershed villages where off-farm income is the important contributor to the household income (Table 10.5).

The availability of long-term data (collected during the years 2001, 2002, 2003, and 2010) on household income in watershed and non-watershed villages confirmed the above facts. It revealed that in watershed villages agriculture and allied activities have provided an opportunity to maximize their revenue through watershed inter- ventions during drought condition. In watershed villages, the income from agricul- ture remained same (around 37%) during drought year (2002), whereas in non-watershed villages, this income has declined drastically from 44% to 12%, but off-farm income grew considerably from 50% to 74% during the same period (Table 10.6). This situation continued even after formal withdrawal of the watershed program in the village. The 10-year long-term data confirmed that the watershed program had positive impacts on rural households in terms of building economic resilience through various in situ and ex situ interventions at the local land- scape levels.

Table 10.6 Impact of integrated watershed management on household income

| Type of villages | Year | Mean household income (US$) from different sources | | | |
		Farm	Livestock	Off-farm	Total
Watershed	2001	308 (36.2)	88 (10.4)	454 (53.4)	850
	2002	202 (36.7)	80 (14.5)	268 (48.7)	550
	2003	492 (46.0)	73 (6.8)	505 (47.2)	1070
	2010	370 (24.0)	98 (6.3)	1076 (69.7)	1544
Non-watershed	2001	254 (43.9)	38 (6.6)	286 (49.5)	578
	2002	50 (12.4)	54 (13.4)	300 (74.3)	404
	2003	388 (42.7)	46 (5.1)	474 (52.2)	908
	2010	166 (16.0)	44 (4.2)	826 (79.7)	1036

Source: Intensive household survey, 2003, 2010

Table 10.7 Distribution of households (HHs) by asset categories in watershed and non-watershed villages

| Asset category (US$) | Watershed | | Non-watershed | |
	% of assets	% of HHs	% of assets	% of HHs
<1280	5.5	20.3	8.9	25.4
1280–1900	9.6	22.0	18.8	30.5
1900–2720	11.1	18.6	17.3	20.3
2720–5140	19.0	18.6	20.6	13.6
>5140	54.8	20.3	34.4	10.2
All HHs	100.0	100.0	100.0	100.0

10.4.3 Household Asset Position

The nature of income generation is likely to vary across socioeconomic classes. In this paper, we take asset ownership as a proxy for socioeconomic class. Table 10.7 shows the distribution of households by the value of the assets owned in each of the two types of villages, viz., watershed and non-watershed villages. The assets here include the value of productive, unproductive, and social assets. Financial assets are excluded from the analysis.

Considering both movable and immovable nature of household assets, households were grouped into different categories based on their asset values. Accordingly, there are five asset categories in our samples. More than one-third of sample households in non-watershed villages have asset value of less than US$ 2000 compared to 44% in watershed villages. On the other hand, only one-fourth of households in non-watershed villages have asset value of more than US$ 2500, whereas in watershed villages, nearly 40% households fall in this group. Therefore, it is clearly revealed that watershed interventions brought opportunities to increase their asset position with enhanced household income and productivity improvement.

Table 10.8   Distribution of households by number of income sources and mean household income

Number of income sources	Watershed villages					Non-watershed villages				
	First	Second	Third	Fourth	All HHs	First	Second	Third	Fourth	All HHs
Distribution of HHs (%)										
Up to two	50.0	50.0	6.3	23.5	30.5	70.6	37.5	14.3	25.0	39.0
Three	16.7	–	37.5	35.3	23.7	29.4	56.3	57.1	25.0	42.4
Four	16.7	42.9	37.5	23.5	30.5	–	6.3	14.3	50.0	15.3
More than five	16.7	7.1	18.8	17.6	15.3	–	–	14.3	–	3.4
Mean income (US$)										
Up to two	500	1098	1899	4403	1677	515	1066	1703	6025	1481
Three	646	–	1719	3471	2317	583	1072	1645	2906	1378
Four	737	1116	1779	4987	2155	–	1021	1534	4147	3219
More than five	607	1212	1555	2847	1737	–	–	2069	–	2069

Source: Intensive household survey, 2010

Table 10.9   Mean measures of income diversification by income quartile

Particulars	Watershed villages					Non-watershed villages				
	First	Second	Third	Fourth	All HH	First	Second	Third	Fourth	All HH
No. of income sources	3.3	3.0	3.9	3.4	3.4	2.0	3.0	3.0	3.0	3.0
Share of off-farm income	0.34	0.31	0.53	0.40	0.40	0.71	0.70	0.69	0.67	0.70
Herfindahl diversification index	0.30	0.30	0.50	0.40	0.40	0.30	0.30	0.40	0.50	0.40
Total income (US$)	581	1114	1722	3937	1984	535	1067	1698	4306	1722

10.4.4   Measures of Income Diversification

As we discussed in the methodology section, we applied three different measures of income diversification to examine the pattern of income diversification in our study area. The analysis showed that about 39% of total households in control villages had merely two income sources, while this proportion was 31% in watershed villages (Table 10.8). However, the proportion of households having more than five income sources was 15.3% in watershed and only 3.4% in control villages. This clearly indicated that the watershed interventions provided more opportunities for rural households to venture into different income-generating activities (Anantha and Wani 2016).

The mean number of income sources was similar across the households (Table 10.9). However, poorer households had few number of income sources compared to richer households both in watershed and control villages. Strikingly, the richer households had more and diversified number of income sources. As discussed in Table 10.9, the share of off-farm income was comparatively high

(68.8%) in control villages. On the other hand, the mean value of HDI was similar for both watershed and control villages, while it varied among poor and rich households. The average total household income in watershed area was higher by 15% compared to control villages. Further, richer households had high average income.

10.4.5 Determinants and Impact of Income Diversification

The preceding sections reveal that most households in the study villages had more than one income source irrespective of their resource endowments. The richer households tend to be more diversified than poor households. However, there was a clear difference between watershed and non-watershed villages in terms of different diversification measures, and that difference was clearly visible in terms of average household income. The results presented in the previous sections are descriptive in nature, and therefore, we have tried establishing their relationship by applying suitable econometric models. By applying two different models, viz., Poisson regression and Tobit regression models, the factors for understanding the pattern of income diversification in the study area were determined. For these models, three different measures of income diversification were considered as dependent variables. The estimated results presented in Table 10.10 showed that all the three measures provided different results except that TLU remained significant in all the models. The TLU was positively related to the number of income sources. On an average, each livestock unit increased the number of sources by 0.1088. This is not surprising as the livestock sector contributed to higher profit, and that would be invested in other subsidiary activities such as petty shops and other micro-enterprises. Our results also showed that there was positive and significant relationship between household productive assets and the number of income sources. Importantly, there was a negative relationship between distances from marketplace and number of income sources. This was obvious given the importance of marketplace for undertaking both skilled and semiskilled jobs for the rural poor. On an average, increase in 1 km distance to access the market reduced the number of income sources by 0.048. In addition, education, farm size, and age of the head of the household are positive in determining the number of income sources but not significant. Importantly, watershed had a positive impact on number of income sources, but there was no significant difference between watershed and non-watershed areas.

In general, share of off-farm income was mainly determined by access to market and level of education. However, in our model none of these variables were significant. Firstly, there was a negative relationship between TLU and share of off-farm income. This clearly suggested that the higher the livestock units, the lower the share of off-farm income because the maintenance of livestock required more time. Therefore, opportunity for undertaking off-farm activities was less. On an average, increase in one livestock unit reduced share of off-farm income by 0.043%. Secondly, there was a negative relationship between farm size and share of off-farm income and road condition. This is not surprising as increased farm size allowed farmers to concentrate on farming activities rather than undertaking off-farm activi-

Table 10.10 Determinants of income diversification

Particulars	No. of income sources (Poisson)		Share of off-farm income (Tobit)		Herfindahl diversification index (HDI) (Tobit)	
	Coefficient	t-value	Coefficient	t-value	Coefficient	t-value
WS_(1=YES;	0.2664266	1.42	−0.29959	−3.85∗	0.0797884	1.44
HH_SIZE_(A	0.0116211	0.40	0.0099513	0.66	−0.0093526	−0.88
HH_SEX_(1=	0.2734502	1.10	−0.0626315	−0.69	0.0481562	0.74
AGE	0.0071584	1.30	0.001372	0.55	0.0006326	0.36
EDUCATION	0.005186	0.32	0.0024533	0.33	0.0140362	2.63∗
FARM_SIZE_	0.0313351	1.40	−0.0196168	−1.67∗∗∗	0.008177	0.98
ASSET_VALU	6.52e-07	1.76∗∗∗	−2.74e-07	−1.44	−1.49E-07	−1.10
ROAD_CONDI	0.0227864	0.19	−0.0956791	−1.76∗∗∗	−0.0181513	−0.47
DISTANCE_F	−0.0487369	−1.87∗∗∗	0.0037538	0.34	−0.0052877	−0.67
TLU	0.1088163	3.40∗	−0.0438806	−2.77∗	0.0376159	3.34∗
CREDIT_(1=	0.0350011	0.26	0.009793	0.16	−0.0035081	−0.08
Constant	0.1988363	0.47	0.7886557	4.47	0.3516757	2.80
Log likelihood	−190.00		−28.299		4.846	

Source: Computed from survey data
∗,∗∗∗ significant at 1 and 10% level; sample size in all models is 118

ties. Since agriculture played a significant role in providing employment and food securities, farmers preferred agriculture over other activities irrespective of their landholding size. Similarly, access to and availability of infrastructure such as roads determined the share of off-farm income. Obviously, bad road condition affected the accessibility of other places (village or market) for daily commuting. Further, it had spillover effects on household income. Surprisingly, the watershed intervention had negative effects on increasing the share of off-farm income. This is true given the importance of agricultural activities in watershed areas due to easy availability of required inputs for undertaking farming. In addition, watershed intervention transformed agriculture into a full-time activity due to increased water availability and sustainable land use planning. However, in non-watershed areas, there was no assured water supply to undertake agriculture in all three seasons. As a result, they ventured into several other off-farm activities to support their livelihood.

The HDI values represent more or less overall diversification scenarios. Surprisingly, in HDI, education had positive and significant relationship with the index value. It clearly suggested that the level of education was a decisive factor for diversifying income activities and a way forward for venture into new skilled and remunerative income opportunities. On the other hand, TLU had positive and significant relationship with HDI. On an average, increase in one livestock unit increased the HDI by 0.037. The livestock sector played a crucial role in enhancing the rural household income as it provided mild drought-insensitive income for rural households.

Since the household variables are endogenous in nature, we applied 2SLS to examine the effects of the three different models on household income (Table 10.11). Considering total household income as a dependent variable and all the three measures along with other household variables as independent variables, farm size had positive association with total household income in all the models. Education was also positively associated with total household income. The share of off-farm

Table 10.11 Impact of income diversification on rural farm household livelihoods

| Particulars | Dependent variable=total household income | | |
	Share of off-farm income	No. of income sources	HDI
SHARE_OF_OFF-FARM INCOME	801.358***		
	(3.825)		
NUMBER_OF_INCOME SOURCES		14065.422***	
		(2.648)	
Herfindahl diversification index			73254.456
			(2.402)**
WS_(1=YES; 0=otherwise)	25694.980	−10540.867	−3775.937
	(1.473)	(−0.603)	(−0.223)
HH_SIZE_(AE)	−450.251	−8.972	428.182
	(−0.147)	(−0.003)	(0.136)
HH_SEX_(1=MALE; 0=OTHERWISE	25970.167	20964.991	18368.193
	(1.359)	(1.064)	(0.934)
AGE (years)	517.028	620.403	608.345
	(0.984)	(1.145)	(1.130)
EDUCATION (years)	4023.046***	4610.436***	4709.497***
	(2.602)	(2.844)	(2.872)
FARM_SIZE_	16814.636***	15581.844***	14999.926***
	(6.726)	(6.114)	(5.956)
ASSET_VALUE	3.747E-02	2.711E-02	2.486E-02
	(0.938)	(0.660)	(0.606)
ROAD_CONDITION	−1712.171	−5026.579	−6899.526
	(−0.149)	(−0.427)	(−0.593)
DISTANCE_F	−1457.047	−1300.385	−565.822
	(−0.625)	(−0.540)	(−0.235)
TLU	5073.649	−979.735	−540.496
	(1.614)	(−0.294)	(−0.166)
CREDIT_(1=yes; 0=otherwise)	7951.316	10384.317	7739.392
	(0.639)	(0.806)	(0.605)
Constant	−64491.779	−38137.775	−29239.344
	(−1.596)	(−0.940)	(−0.735)
R^2	0.54	0.51	0.514
Adj R^2	0.49	0.45	0.459
F statistics	10.276	9.064	9.272

Figures in parentheses are "t" values; ***, ** significant at 1 and 5% level

income on household income at 1 percentage point had an effect of Rs. 801. Similarly, each additional income source provided additional annual income of Rs. 14,065. Obviously, off-farm activities were more lucrative than farming alone, so diversification was pursued as a strategy to increase household income, whenever the opportunity had arisen. Yet farm size had a significant effect too in all the three models. This suggested that while off-farm activities increased income, farming still remained important for household livelihoods in rural areas.

10.4.6   Changing Scenarios of Income Diversification

An attempt was made in this section to examine whether any changing trend in the study area correlated with that of the national scenario in terms of income diversification. We have compared contribution of different income sources with the results of the recent NABARD All India Financial Inclusion Survey 2016–2017 using recent household survey data. The survey launched in the beginning of 2018 in Kothapally involved detailed enquiry into the amount of household income from various sources in the last 1 year preceding the survey. The net income for households was derived by adding income from all sources for a particular household and deducting the expense incurred toward pursuing income-generating activities like cultivation, livestock rearing, and other enterprises.

The survey results show that agriculture is the main source of income in Kothapally as it contributes more than 50% to total household income compared to only 19% at rural areas of national level (Table 10.12). It is to be noted that there is difference between Kothapally and India in terms of households with land and those who are landless. In Kothapally, households with land are far higher (98.5%) as compared to 48% in India and 47% in Telangana state. Secondly, farm income from farm households is 35% at all India level. The primary source of income in rural areas is from wage labor which includes agricultural labor and skilled and non-skilled labor work. Even in Kothapally, wage labor contributes about 22% which is also a significant source of income for rural households. Besides, the live-

Table 10.12 Average monthly household income by sources in Kothapally and India

Sources of income	Kothapally[a]	All India[b]
Cultivation	6171 (53)	1494 (19)
Livestock rearing	1572 (14)	338 (4)
Wage labor	2489 (22)	3504 (43)
Govt./pvt. services	702 (6)	1906 (24)
Petty business/other enterprises	110 (1)	679 (8)
Other sources	517 (4)	138 (2)
All sources	11,561 (100)	8059 (100)

Source: [a]Primary survey carried out during 2018; [b]NABARD All India Financial Inclusion Survey 2016–2017

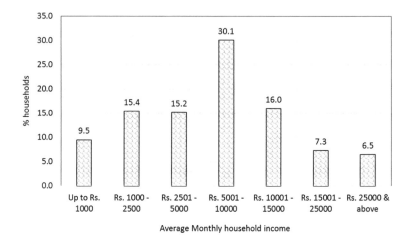

Fig. 10.1 Distribution of households by monthly household income in Kothapally, Telangana, India

stock sector plays an important role in rural areas as it acts as insurance during drought situations. In Kothapally, income from livestock rearing contributes around 14%. With the changing educational status and living standards, more and more number of people are engaged in public and private sector employment which is contributing about 24% to total household income at the national level. However, it is about 6% in Kothapally as more people are engaged in low-profile jobs. This indicates that Kothapally is largely an agrarian village with more number of farmers pursuing farming as an enterprise with the availability of water and market facilities.

Figure 10.1 describes the distribution of monthly household income in Kothapally. It was found that about 70% of the households in the village are earning about Rs. 10,000 per month, whereas the remaining 30% of the households earn more than Rs. 10,000 per month. The study findings corroborate with NABARD's All India Financial Inclusion Survey findings (NABARD 2018).

10.5   Conclusion

Analysis of income diversification among farm households in rural Telangana (erstwhile Andhra Pradesh), India, revealed that irrespective of areas (i.e., watershed or non-watershed), households are diversifying their income sources. However, agriculture remained a mainstay for rural farm households both in watershed and non-watershed areas. Agriculture contributed more than 50% in watershed area, while this proportion was less than 50% in non-watershed area. This clearly suggested that the watershed interventions provided more favorable opportunities to farm households to undertake full-time farming, whereas non-watershed areas lacked these facilities. Further, the share of off-farm income in total household income was highest in non-watershed areas compared to watershed area. Within off-farm sources, wage earning (casual and agricultural) contributed the largest share followed by business and service activities.

Minimum literacy, better infrastructure, and nearest market availability were identified as important drivers for diversification of income for farm households in rural areas. However, farm size also played a crucial role in diversification in watershed areas due to its increasing intrinsic value because of various in situ and ex situ interventions.

The policy implications of these findings rest with enhancing the household's access to off-farm activities, which support equitable rural development. This requires development of not only the physical infrastructure such as roads, electricity, water works, and telecommunications but also improvement in rural education and financial markets. As evident from these findings, farming emerged as one of the major income sources in watershed areas. Hence, attempts need to be done on utilizing farm products locally for promoting off-farm activities. These interventions will help farmers to find local market for their products as well as to meet day-to-day expenditure.

With watershed development, Kothapally attracted labor force from surrounding villages which if scaled-up definitely will reduce demand for MGNREGA activities and empower rural people. Due to increased water availability, the higher share of income from farming has increased dramatically over baseline which is in tune with the increased cropping intensity of around 140% as compared to 120% at all India level. Further, for achieving the SDG of zero hunger and reducing poverty as well as building climate resilience, the integrated watershed development model is the best approach as demonstrated by the Kothapally model over the last 20 years.

References

Anantha, K. H., & Wani, S. P. (2016). Evaluation of cropping activities in the Adarsha watershed project, southern India. *Food Security, 8*(5), 885–897.

Babatunde, R. O., & Qaim, M. (2009). Patterns of income diversification in rural Nigeria: Determinants and impacts. *Quarterly Journal of International Agriculture, 48*(4), 305–320.

Barnett, C. B., & Reardon, T. (2000). *Asset, activity and income diversification among African agriculturists, some practical issues.* Unpublished project report to the USAID BASIS CRSP, Washington DC, USA.

Ersado, L. (2005). Income diversification before and after economic shocks: Evidence from urban and rural Zimbabwe. *Development Southern Africa, 22*(1), 27–45.

FAO (Food and Agriculture Organisation). (2006). *The new generation of watershed management programmes and projects* (FAO Forestry Paper 150). Rome: FAO.

Garg, K. K., et al. (2012). Assessing impact of agricultural water interventions at the Kothapally watershed, Southern India. *Hydrological Processes, 26*(3), 387–404.

GoI (Government of India). (2007). *Eleventh five year plan*. New Delhi: Government of India.
GoI (Government of India). (2008). *Common guidelines for watershed development projects*. Department of Land Resources, Ministry of Rural Development, Government of India.
GoI (Government of India). (2011). *Census of India – 2011*. New Delhi: Office of the Registrar General and Census Commissioner, Ministry of Home Affairs, Government of India.
Joshi, P. K., Jha, A. K., Wani, S. P., Joshi, L., & Shiyani, R. L. (2005). *Meta analysis to assess impact of watershed program and people's participation. Comprehensive assessment research report 8. Comprehensive assessment secretariat*. Colombo: International Water Management Institute (IWMI).
Joshi, P. K., et al. (2008). *Impact of watershed program and conditions for success: A meta-analysis approach (Report no. 46)*. Patancheru: International Crops Research Institute for the Semi-Arid Tropics.
Joshi, P. K., et al. (2009). Scaling-out community watershed management for multiple benefits in rainfed areas. In S. P. Wani, J. Rockström, & T. Oweis (Eds.), *Rainfed agriculture: Unlocking the potential* (pp. 276–291). Wallingford: CAB International.
Kerr, J., et al. (2000). *An evaluation of dryland watershed development in India* (EPTD discussion paper 68). Washington, DC: International Food Policy Research Institute.
Minot, N., et al. (2006). *Income diversification and poverty in the Northern upland of Vietnam* (Research report no. 145). Washington, DC: International Food Policy Research Institute.
National Bank for Agriculture and Rural Development. (2018). *NABARD all India rural financial inclusion survey 2016–2017*. Mumbai: National Bank for Agriculture & Rural Development.
Palanisami, K., et al. (2009). Evaluation of watershed development programmes in India using economic surplus method. *Agricultural Economics Research Review, 22*, 197–207.
Pathak, P., et al. (2007). *Rural prosperity through integrated watershed management: A case study of Gokulapura-Goverdhanapura in Eastern Rajasthan* (Global Theme on Agroecosystems Report No. 36). Patancheru: International Crops Research Institute for the Semi-Arid Tropics.
Raju, K. V., et al. (2008). *Guidelines for planning and implementation of watershed development program in India: A review* (Global theme on agroecosystems report no. 48). Patancheru: International Crops Research Institute for the Semi-Arid Tropics. 95 pp.
Reardon, T., & Vosti, S. A. (1995). Links between rural poverty and the environment in developing countries: Asset categories and investment poverty. *World Development, 23*, 1495–1506.
Shiferaw, B., Anupama, G. V., Nageswara Rao, G. D., & Wani, S. P. (2002). *Socioeconomic characterization and analysis of resource use patterns in community watersheds in semi-arid India.*. Working paper Series No. 12. Patancheru 502324 (Vol. 44). Andhra Pradesh, India: International Crops Research Institute for the Semi-Arid Tropics.
Sreedevi, T. K., et al. (2006). *On-site and off-site impact of watershed development: A case study of Rajasamadhiyala, Gujarat, India* (Global theme on agroecosystems report no. 20). Patancheru: International Crops Research Institute for the Semi-Arid Tropics.
Sreedevi, T. K., et al. (2008). *Impact of watershed development in low rainfall region of Maharashtra: A case study of Shekta watershed* (Global theme on agroecosystems report no. 49). Patancheru: International Crops Research Institute for the Semi-Arid Tropics. 52 pp.
Wani, S. P., et al. (2003). *Farmer-participatory integrated watershed management: Adarsha watershed, Kothapally, India: An innovative and upscalable approach*. Patancheru: International Crop Research Institute for the Semi-Arid Tropics.
Wani, S. P., et al. (2012). Enhancing agricultural productivity and rural incomes through sustainable use of natural resources in the semi-arid tropics. *Journal of the Science of Food and Agriculture, 92*, 1054–1063.
Wani, S. P. et al. (2006). Issues, concepts, approaches and practices in the integrated watershed management: Experience and lessons from Asia. In: *Integrated management of watershed for agricultural diversification and sustainable livelihoods in Eastern and Central Africa: Lessons and experiences from Semi-arid South Asia*. Proceedings of the international workshop held 6–7 December 2004 at Nairobi, Kenya, pp. 17–36.

Wani, S. P., et al. (2008). *Community watershed as a growth engine for development of dry-land areas: A comprehensive assessment of watershed programs in India* (Global Theme on Agroecosystems Report No. 47). Patancheru: International Crops Research Institute for the Semi-Arid Tropics. 156 pp.

Wani, S. P., et al. (2011). Assessing the environmental benefits of watershed development: Evidence from the Indian semi-arid tropics. *Journal of Sustainable Watershed Science and Management, 1*, 10–20.

Chapter 11
Robust Rural Institutions and Governance Are Must for Sustainable Growth in Watersheds

K. V. Raju and D. S. Prasad Rao

Abstract This chapter focuses on the type of institutions existing and functional over the years in this *gram panchayat* (village council) and also discusses how they are governed as organizations; both formal and informal institutions are discussed here. The governance mechanism evolved over the years, its refinements made over time, and their structures for governance are analyzed. Authors have critically analyzed the institutional mechanism; the management process; key organizations and their role, including the pivotal role played by the local village council; and how local women groups through their self-help groups successfully dealt with microfinancing and enabled convergence with other programs. Also described are the utilization of spent malt to boost up milk production, how water rights were tactfully managed with support from a local non-governmental organization, and methods followed by watershed user committees for revitalization of water bodies.

Keywords Rural institutions · Governance · Watersheds · Village council · Insitutional mechanism · Micro-finance · Self help groups

11.1 Introduction

The governance mechanism evolved over the years, its refinements made over time, and their structures for governance are analyzed in this chapter. The literature related to Kothapally, available till recently, has rarely focused on rural institutions and their governance; over the last 17 years (2000–2017), several publications, based on both desk review and field survey, were published (CGIAR 2003). Its review indicates scant attention on institutions and governance mechanism. Thereby,

K. V. Raju (✉)
Former Theme Leader, Policy and Impact, Research Program-Asia, International Crops Research Institute for the Semi-Arid Tropics (ICRISAT), Hyderabad, Telangana, India

D. S. Prasad Rao
Research Program Asia, International Crops Research Institute for the Semi-Arid Tropics (ICRISAT), Hyderabad, Telangana, India

© Springer Nature Switzerland AG 2020
S. P. Wani, K. V. Raju (eds.), *Community and Climate Resilience in the Semi-Arid Tropics*, https://doi.org/10.1007/978-3-030-29918-7_11

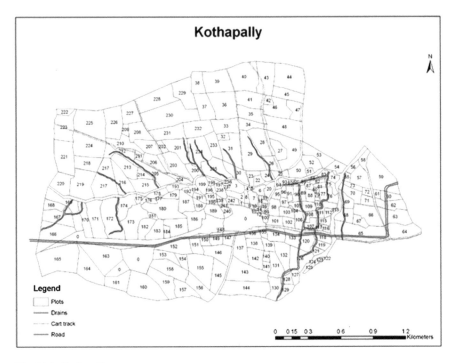

Fig. 11.1 Kothapally map

this chapter emphasizes on this dimension, based on extensive discussions with stakeholders and walkthrough survey in farmer fields.

In the course of mapping the rural organizations and institutions in Kothapally, we tracked (a) *gram panchayat*, (b) women self-help groups (SHGs), (c) watershed user committees, (d) watershed association, (e) farmers' association, (f) local NGO, and (g) water rights and water distribution mechanism. The authors held a series of discussions with various stakeholders, both on one-to-one and in a group, spread over several weeks in Kothapally village. Additionally, they made a content analysis of the records maintained by the *gram panchayat*, SHGs, and watershed user committees (Fig. 11.1).

11.2   Institutions and Governance

India has introduced strong legal measures for conservation of natural resources. Several innovative approaches are in place to ensure people's participation in conservation. However, the available indicators on the progress of the implementation of these measures show that there are several gaps and weaknesses in the conservation and management[1] (Raju et al. 2014). In India, the government through its com-

---

[1] Raju, K.V., S. P. Wani, Poornima, S. 2014. Institutions for Ecosystems Services in Semi-Arid Tropics: A Study of a Cluster of Villages in South India (unpublished paper).

mand and control methods alone cannot manage the natural resource conservation successfully, and this calls for looking at strengthening alternate options, particularly economic instruments.

11.2.1 Institutional Mechanism

Institutions that properly coordinate the human use of ecosystems have several functions. They must create management structures that promote the definition of multiple objectives and coordinate organizational tasks in a cost-effective way. They also must create management processes that are legitimate and flexible and promote socially appropriate time horizons that recognize intergenerational rights to resource use. Each of these functions has economic dimensions that are critical to their successful implementation (Hanna1998).

For any ecosystem to be used in a sustainable manner under management practice, sets of institutions and property rights regimes are required that reflect the attributes of the ecosystem and its human users that value ecosystem services as well as ecosystem commodities on a broad ecosystem scale.

In this context, the word "institutions" refers to the different formal and informal groups in the Kothapally village that are acting as a body using the ecosystem services and depending on ecosystem services for agricultural activity and also groups that are working for maintaining and conserving the natural ecosystem of the region (along with enriching knowledge of its importance in the region).

Institutions or stakeholders involved in ecosystem maintenance in the region are categorized into two groups, formal and informal institutions. Formal organizations include Telangana state *gram panchayat* (village council), Stree Nidhi. Informal groups/institutions include village leaders, women self-help groups, farmers' groups, and youth groups. Also reviewed are local governance mechanism and rules and regulations to protect local natural resources.

Property rights regimes are a subcategory of institutions, bundles of entitlements that define owners' rights and duties, and the rules under which those rights and duties are exercised (Bromley 1991). As, in this semi-arid region, few resources like water, soil fertility, and groundwater potential are owned by or belong to the owner of that land and not as a natural resource of the area, this caused misuse or overuse of the resource by an individual which is further affecting the whole region. This is also due to poor or minimal regulations and policies for natural resources usage in individual-owned land (e.g., agricultural land). A limitation in the property rights regime is that they do not specify claims to the full edge of goods and services provided by an ecosystem. The lack of full specification means that it is unclear who can claim rights of use or how those rights may be used. This has to be clear in case of making policy as at present all resources in a patch of land belong to the owner in an undeclared way.

11.2.2   Management Process

All the formal and informal institutions in rural ecology are following their own rules and regulations. No one is working in coordination with other institutions. Most of the government bodies work on their own, which led to unsustainable implementation of conservation plans for conserving the ecosystem.

In the management process, three components are particularly influenced by economics: the time horizon over which management takes place, the legitimacy of the management process, and the flexibility with which management takes place.

Time horizons: Environmental resources are a natural form of capital that has value in both the size of the stock and the flow of services. Sustaining ecosystem requires constraining human exploitation at the levels that will ensure the continuance of the stock and flow of ecosystem benefits into the indefinite future. The time horizon over which ecosystems are managed is critical to their rates of use. Future expected benefits from any resource have less current users and managers that do know current benefits and so the future is discounted relative to the present. The shorter the time horizon over which planning is done, the greater the discounting of the future benefits and higher the rates of current use (1998, Hanna) for purpose of sustainability and intergenerational equity; it is in social interest to manage time horizon as far as to the future as possible using very low rates of discount. In Kothapally *gram panchayat* also, this social interest has to be highlighted in order to run institutions in a sustainable way for resource management and use at present and also conserving them for future agricultural practices; otherwise, soon in the next few decades, the whole village may be turned into an urban setup with already lost interest in agriculture and increasing land value for building private companies and industry as the area is under the attack of urban sprawl area of a metropolitan city.

Legitimacy: To be a legitimate management organization, procedures by which it functions, the content of its regulations, the method of enforcement, and the distribution of its outcome must all be perceived to be fair. Conflict between resource users is inevitable and rent-seeking is widespread, so procedural legitimacy also requires a transparent mechanism for conflict resolution. As in the case of this Kothapally village, there is high difference between each institution in terms of social responsibilities and economic status, but in a village like so, everything put aside should be looked at equally, and resource management and resource benefits also need to be shared equally which is lacking in the present situation.

Flexibility: In several dimensions, requirements for adoptability to promote ecosystem resilience are in opposition to requirements for reducing uncertainty and promoting efficiency and legitimacy. In theory the ideal institution exhibits efficiency (North 1992) by accommodating an ability to respond to change. Because government institutions with policies and regulations to conserve resources do not change with time, they have less ability to respond to a sudden change and have to be looked after; for this purpose, in this study area we can see that the private sector takes the responsibility of doing research and development based on changing people's perception over time in which all educational institutions come into picture like ICRISAT, to understand the status of resource use and dependency on ecosystem services in a time period of 30 years and then decide what is best to bring in the change to sustainable use of resources in the region.

11.3   Key Organizations and Their Role

11.3.1   Kothapally Gram Panchayat

In Telangana state, out of the total 8864 *gram panchayats*, 91 are headed by women; Smt. Shobha rank her *gram panchayat* in the top 10 category, in terms of overall performance. Also in terms of setting up women SHGs and achieving 100% hospital delivery for pregnant women, 100% cleanliness, and 100% open defecation-free (ODF), all of these happened under the dynamic leadership of the *gram panchayat* president (*sarpanch*) Smt. Shobha Sudhakar Reddy (44 years old, B.A. second year), president for the last 4.5 years. She got elected as the first woman president under the women's reservation category, during the elections held in 2013. For the same reason, she was invited to represent the Telangana state in several national and regional conferences and workshops held across India and she is very proud of it. Even majority of the villagers feel very proud of her performance. Thereby, both the *gram panchayat* elected members and the several informal associations like caste groups and youth associations support her several initiatives for the development of the village. Some of them are discussed in the later sections of this chapter (Fig. 11.2).

The gram panchayat is crystal clear in describing its role in enabling ICRISAT initiatives for the development of this village. She said "The *gram panchayat* facilitated several rounds of discussions between the ICRISAT scientists and the farmers. It also provided its office space for both formal and informal discussions over the years. On the other hand, it seeks opinions of various categories of people from this village and aggregates it before sharing with the ICRISAT group."

Fig. 11.2 Smt. Shobha Sudhakar Reddy, president, gram panchayat, Kothapally

How about handling disagreements at certain levels? The president confirms that "Whenever there are disagreements I or my council members personally talk to those individuals and in a couple of days we generally reach an amicable settlement. This may be related to locations of watershed structures, sharing of work among user group members, sharing of irrigation water and other benefits." She also said that "If there is any issue or problem within the village they will not go to the local police, but it will be solved by village committee only."

Have your efforts brought any benefits to the village? Smt. Shobha Reddy elaborated her happiness in accrued benefits owing to ICRISAT intervention; she said "Look at this now, our village has become world famous; visitors from various countries and senior officers from the Government of India and the state government keep visiting us as a model village. A large number of neighboring villagers keep visiting our farmers' plots; filled-up wells, continuously pumping out bore well water, and green farms even during summer have surprised them. Many farmers and occasional visitors have personally told me and my council members about the change they are witnessing over the decades from dryness to greenness. Recently the trainees from the Indian Administrative Service selected this village for a 3-day stay and walked around the village and the farm plots to understand the development process. We feel proud of this new 'wow attitude' towards us."

What else had surprised you? The president thought for a while and responded, "Recently a senior journalist from the state capital Hyderabad from a well-known newspaper walked for half a day around the watershed structures and farm plots and at the end gave us a surprise visit to my office. The journalist said it's unbelievable to see the greenness in the village, during the summer, while the neighboring villages are all dry; he appreciated us and then wrote half a page news story on our village" (see Box 11.1).

Box 11.1: News Story on Kothapally Village in *Namasthe Telangana*, July 24, 2017

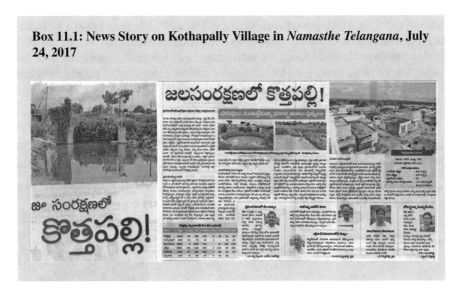

Encouraged by these developments and supported by council members and villagers, the president has gone ahead with a few more activities; she got the open drains of the streets, infested with insects and having stench/bad smell, cleaned up and covered with slabs. Also, she got them restructured and connected to the wastewater treatment plant installed at the end of these drains. It was designed, constructed, and operated and then transferred to the village by the ICRISAT scientists. "This has become another attraction in our village," she said. Indeed, recalling the process in setting up a wastewater treatment plant, the president said that the *gram panchayat* faced tough questions from the nearby farmers on sharing their area for wastewater plant construction and its location. It took several days to discuss individually with each household and with farm owners to support this initiative. Finally, after several rounds of persuading and explaining the benefits of wastewater recycling, the local residents agreed. But again, when the crops were grown by using the recycled water, the farmers were skeptical about the quality of the crops, e.g., maize grown by using the recycled wastewater. The president volunteered to eat that maize crop in front of a bunch of residents to prove that the maize has no bad elements; this had convinced several farmers to start using the recycled wastewater.

Safe Drinking Water Supplies Similarly, while setting up the reverse osmosis (RO) plant for providing clean and adequate drinking water to households, there was funds crunch. The gram panchayat had a round of discussions with the residents and collected Rs 200 to Rs 500 per household. This initial money enabled ICRISAT to mobilize further funding to install the RO plant. Today all households utilize this on pay-per-use basis; a 20 l can cost Rs 3, paid through prepaid cards by swiping the water vending machine. Daily 100 cans of 20 l each are consumed; each 20 l can cost Rs 3. From the amount collected against the cans consumed, Mr. Ramesh, who is the caretaker of the plant, is paid Rs 300 per month, Rs 500 is spent on other chemical needs for the plant, Rs 500 is spent to change the filter, and the balance amount is retained by the panchayat committee for future maintenance of the plant, if any (Fig. 11.3).

Over the years the *gram panchayat* has learned the art of organizing public meetings. Steering the discussions on a positive note, during general functions like temple festivals, the *gram panchayat* council coordinates and allocates the functions on caste group basis: based on their skills and the group size. Also there is a separate

Fig. 11.3 RO plant in Kothapally village

work for the youth which involves regulating the seating arrangement for a large gathering and restricting the usage of mobile phones. Interestingly, the gram panchayat uses WhatsApp group messaging for faster and wider communication. Such a systematically organized public function was held in the village after 26 years, as several villagers opined during our discussions. This has enhanced the confidence level of the general public on the organizational and functional skills of the *gram panchayat* president and its members.

Being a woman president of the *gram panchayat,* she has proven that an educated woman can do much better than a man, owing to her having more patience and persuasion skills. Incidentally, she said this, just a few minutes before her departure for a women's day function on March 8, 2018.

Do you think the symbiotic relationship between the residents of the village and the *gram panchayat* will continue in future years? She pondered over this question and recalled her discussions with the presidents (both male and female) in neighboring villages. She said "Increasingly, I observe the dependency of individuals on *gram panchayat* is declining. Mainly because individual's economic status improved and their linkages to the nearby urban areas and information technology have enormously increased; however to be people's president in the *gram panchayat,* it is necessary that the president (man or woman) should regularly interact with a large number of people and develop serving attitude."

11.3.2 Stree Nidhi

Kothapally village has a total of 40 SHGs, as per the guidelines of the Telangana state Stree Nidhi (an apex body at state level). At the village level, there is one *gram sangh* (village federation). This village association works as an apex body for all the SHGs within the village. The loan amount from the bank is routed through this village association. Similarly, the individual SHG monthly installments are collected by this village association and paid to the bank. All the records and bookkeeping are handled by this village association. For bookkeeping a person (cluster coordinator) was appointed by the bank through consultation and approval by these SHGs and village association. Over the last 2 years the local cluster coordinator seems to have improperly handled the SHGs' funds. The gram panchayat president, herself a member in one of the SHGs, hopes to set this process right very soon. This village association has two members from each individual SHG, thereby a total of 80 members in the village association.

The village association, besides handling the microfinance, also is closely involved in several women and children welfare activities in the interest of the larger society. Some of them are (a) stopping child marriages, (b) rehabilitating child labor and sending them to school, (c) closing alcohol-selling shops within village premises, (d) construction of toilets in all households, and (e) providing awareness and orientation in communication both within the association in handling meetings and corresponding with bank and external agencies.

On the other hand, the small SHGs (a) deal with problems of women members and escalate them to village association, (b) conduct two meetings per month on microfinance and women issues, (c) deal with the non-repayment aspects, and (d) handle women- and children-related issues.

The village association mobilized a loan of Rs 200,000 in 2006 and Rs 170,000 in 2008 and Rs 1,000,000 from Stree Nidhi cooperative in 2017. Thus a total of Rs 1,370,000 has been channeled through 40 SHGs to 400 households. A household may have many women as members of an SHG. But each woman of that household should be a member of a different SHG.

This village association on an average had lent a loan of Rs 25,000 per household across all 400 households in the village. Earlier loan per household was restricted to Rs 10,000; gradually it increased to Rs 25,000. At present a woman repays Rs 1210 per month. This process has reduced women's social insecurity and enormous dependency on money lenders who used to charge high interest rates of 24–48% over the year. Today all these women feel enormously liberalized from the money lenders' clutches and feel empowered in mobilizing a loan in a very easy process and are also able to repay in a more convenient manner with less interest rate of 12%.

The village association has got some Rs 6000 as earning through interest, which is considered as their own fund. Even this money has been lent out to the needy women at 1% interest. The village association now has their own building constructed at a cost of Rs 150,000 supported by the Government of Telangana. The building was constructed by three women groups, who had taken the construction activity as contractors, and each of them is able to make a profit of Rs 5000 in this process. In the future some 150 women, all members of SHGs, are willing to work in a group for any economic activity related to micro-enterprises or even farm labor work.

NABARD launched the SHG bank linkage program in the country 25 years ago, which ushered in a great revolution and has been a game changer in the sphere of finance including those unreached. As per the 2016–2017 NABARD report on the status of the microfinance sector, there are 85.7 lakh SHGs in the country covering around 10 cr rural households penetrating 60% of the rural households. The loan amount disbursed by banks to SHGs during the year 2016–2017 was Rs 38,781.16 crore, while the total bank loan outstanding to SHGs was at Rs 61,581.30 crore. The average loan disbursement per group in the country in that year was 2.04 lakh. Microfinance is instrumental in achieving Sustainable Development Goals of poverty eradication by 2030 through financial inclusion and social engineering.

An appropriate ecosystem is being created by GOI through NRLM in this direction by organizing and nurturing SHGs and their federations more effectively. In development economics, empowerment of women in all spheres of life assumes great importance, and keeping in view SDGs, it is imperative that such a network of SHGs and their federations are properly harnessed for planning development of the state as SHGs and their federations serve as entry point for many government programs for ameliorating conditions of poor women.

11.3.2.1 Stree Nidhi: An Appropriate Solution

Inclusive development began in 1992 with the launch of the SHG bank linkage program in the country under the aegis of NABARD. Formation of SHGs and their linkage with banks have expanded credit flow for consumption and other productive purposes tremendously. However, due to various factors, the poor is not able to access adequate and timely credit. They are, therefore, resorting to high-cost credit provided either by microfinance institutions/other leaders in the private sector and/or money lenders. This has pushed the poor into a debt trap and some of them committed suicide in the erstwhile state of Andhra Pradesh. Then the state government promulgated an ordinance in the year 2010 regulating the operations of MFIs in the state. While there was huge demand for credit, the supply was not adequate from the banking sector.

The state government established Stree Nidhi Credit Cooperative Federation Ltd., in association with Mandal Samakhyas and Pattana Samakhyas and registered on September 7, 2001, under the State Co-operative Societies Act of 1964 (Fig. 11.4).

The main objectives of Stree Nidhi are (a) to provide affordable credit to SHG members expeditiously using technology and supplement credit support from banking sector, (b) to alleviate poverty by financing income-generating activities, and (c) to work for socioeconomic upliftment of members of self-help groups both in rural and urban areas.

Over a period of the last 6 years, Stree Nidhi has exhibited robust growth and emerged as a major boon to the poor in rural and urban areas across the state. In the present scenario, the poor is not required to go anywhere except Stree Nidhi for sourcing credit both for the emergent and livelihood requirements as credit is available all the time and at a low cost. The following features shown in Fig. 11.5 distinguish Stree Nidhi from others in the microfinance sector.

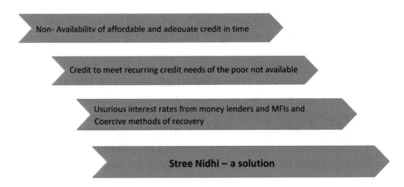

Fig. 11.4 Emergence of Stree Nidhi. (Source: Government of Telangana2018. Stree Nidhi, Annual Report 2016–2017)

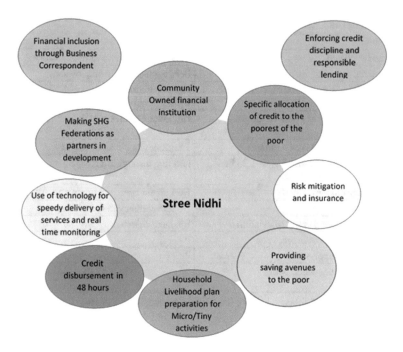

Fig. 11.5 Unique features of Stree Nidhi. (Source: Government of Telangana, 2018 Stree Nidhi, Annual Report 2016–2017)

11.3.2.2   Providing Credit at an Affordable Cost

The interest rate charged to SHG members on loans is presently at 13.0% per annum, and effective January 1, 2018, it is 12.5 only. Of this 1.25% is passed on to VOs/SLFs and 0.25% to MSs/TLFs to compensate for their services in monitoring Stree Nidhi services. Thus, the effective rate of interest to Stree Nidhi is less than that of banks in the state and may be still lower than the banks if the real cost of transaction and other costs of availing loans from banks are taken into consideration under the SHG bank linkage program. The interest rates charged by major banks in the state under the SHG bank linkage program are mentioned below (Table 11.1).

Unlike the rate charged by MFIs in the private sector which is at 20–21%, the effective interest charged presently by Stree Nidhi is at 11.5%, which is further reduced to 11% from January 1, 2018.

Table 11.1 Interest rates charged by banks under the SHG bank linkage program

Name of the bank	Rate of interest (%)
State Bank of India	12.25
Andhra Bank	11.90
APGVB	12.50
TGB	14.00

Source: Government of Telangana 2018. Stree Nidhi, Annual Report 2016–2017

Stree Nidhi vs Other MFIs	
Stree Nidhi	Other MFIs
Community owned	Promoted by private individuals/organizations
SHG federations – last-mile connectivity	Lending using only employees
Need-based lending	High operational costs at 8–10% of the working capital
Low operational cost at 1–2% of the working capital	High interest rates – 21%
Low interest rates – 13% to members	1% processing charges levied
margin shared with VO and MS @ 1.255 and 0.25% each	Lending through JLGs
No processing charges levied	No holistic approach to poverty alleviation
No profit motive	No transparency in functioning
Holistic approach to poverty alleviation	Business oriented
No unhealthy practices in recovery	
Transparency in functioning	
Concern for the poor while being self-sustainable	

Source: Government of Telangana 2018. Stree Nidhi, Annual Report 2016–2017

11.3.2.3   Convergence with SERP and MEPMA

Convergence with SERP and MEPMA brings better synergy and more value to the services of Stree Nidhi. While SERP and MEPRM add value through strong institutions, financial access, livelihood promotion, and social interventions, Stree Nidhi can meet investment and other needs of the community in addition to extending banking services and government-to-citizen services by functioning as a business correspondent to banks (Box 11.2).

Box 11.2: Stree Nidhi as a National Support Organization (NSO)

Considering the utility of Stree Nidhi and need for replication in other states, NRLM, Government of India, has identified Stree Nidhi, Telangana, as a national support organization. The objective is to extend support to SRLMs of different states to replicate the Stree Nidhi model.

Upon the request of SRLMs of Rajasthan and West Bengal, Stree Nidhi has completed the task of preparing feasibility report and DPR for Rajasthan and feasibility report for West Bengal and forwarded these to the respective state governments. Other states, namely, Tamil Nadu and Madhya Pradesh, have also evinced interest in rolling out the Stree Nidhi model in their states.

Source: Government of Telangana 2018. Stree Nidhi, Annual Report 2016–2017

11.3.2.4   Microcredit to Achieve Macro-goals

Stree Nidhi Credit Co-op Low-Cost Credit Delivery

Microfinance, having been pioneered in Bangladesh, would find its zenith in Andhra Pradesh for a while as credit levels surged. It was too good to last as things went haywire after reports of suicides following people being coerced into repaying loans, prompting the state to step in. Repayments dwindled. MFIs had to fold up and tiny loans dried up. However, a turnaround has taken place in the last few years. Sumalata, now in her 30s, for instance, recently got a loan to set up a dairy.

Before the 2001 crackdown, Andhra Pradesh had the highest concentration of microfinance operations with 17.3 million SHG members and 6.2 million MFI borrowers. Total microfinance loans including both SHGs and MFIs stood at Rs 15,769 crores with average loan outstanding per poor household at Rs 62,527, the highest among all the states. Andhra Pradesh accounted for 62% of total MFI business in India before the squeeze. The passing of the Andhra Pradesh Microfinancing Institutions (Regulation of Money Lending) Act in December 2010 imposed stringent rules on MFIs operating in the state, bringing the sector to a halt. With the sudden void in the fund flow, much of the rural entrepreneurial spark was extinguished. Many were forced to sell homes, vehicles, and cattle to make ends meet. Others cut down on businesses amid shrinking earnings.

According to a joint study by MicroSave and IFMR published in June 2012, 59% of the credit needs in undivided Andhra Pradesh were met by money lenders in the absence of MFIs. But from then on, the story took a positive turn when the Stress Nidhi Credit Co-operative Federation, promoted by the Andhra Pradesh government and the federations of SHGs, entered the scene. Credit flow started again, and banks too started giving money to SHGs. Over the years, Stree Nidhi has created a niche in the sphere of microfinance with its low-cost credit delivery.

The organization has so far extended credit support to some 1.66 million borrowers at 13.5% on a reducing balance. There is no processing fee for taking a loan. Bharat Financial Inclusion, which is the least-cost provider of microloans among microfinance companies, charges 19.75% a year. There are also underlying resilience and unity that resonate in this turnaround story. The Society for Elimination of Rural Poverty (SERP) promoted by the state government has nurtured about 441,000 SHGs in the state's rural belt. Under its program, about 20–30 SHGs in a village come together to form an organization that plays a key role in providing last-mile connectivity, facilitating members to avail themselves of the services of Stree Nidhi.

"Bank loans are not sufficient," she said, echoing the common refrain of rural borrowers across India. "WE get bank loans at 11.5% but then you have to pay processing fee and documentation charges in addition." With encouragement from Stree Nidhi, she has also set herself up as a business correspondent for the Central Bank of India, which has its nearest rural branch 3 km away.

When Andhra Pradesh was bifurcated in 2014, Stree Nidhi was also divided into two. The success of Stree Nidhi Telangana in delivering low-cost funds to borrowers in need has attracted national attention. The National Rural Livelihood Mission has tasked the organization to replicate its model in other states willing to adopt it. Rajasthan and West Bengal have invited Stree Nidhi to conduct feasibility studies. "The emergence of Stree Nidhi as a community-owned, low-cost microfinance provider was a kind of historical necessity," Mohanaiah said. This has brought back justice for poor borrowers by providing them timely and affordable credit. However, it needs to go a long way to eliminate poverty by promoting livelihoods and checkmate money lenders effectively.

This success notwithstanding, money lenders still exist to fill the institutional credit gap. The Telangana government itself said in 2017 in its socioeconomic outlook report that money lenders play a dominant role in meeting the credit needs of 50.6% of households in the state and 51% of total credit disbursed (Figs. 11.6 and 11.7).

11.3.3 Women Self-Help Groups

During the late 1990s, while the watershed structures were under construction, the majority of the households used to borrow money from the local money lenders at high interest rates of 24%, 28%, and up to 36% per year. This was a serious constraint, even for small household needs; people had to depend on these money lenders as they had no other option. With a lot of difficulties, two women SHGs

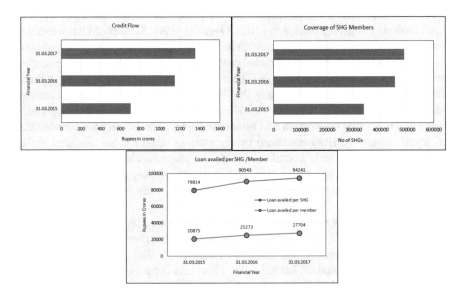

Fig. 11.6 Stree Nidhi performance highlights. (Data Source: Government of Telangana 2018. Stree Nidhi, Annual Report 2016–2017)

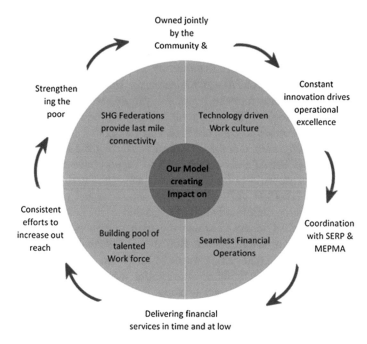

Fig. 11.7 Stree Nidhi model. (Source: Government of Telangana 2018. Stree Nidhi, Annual Report 2016–2017)

were started from the support provided by the state government. Realizing the benefits of these SHGs, more women had formed additional SHGs over the years. The ICRISAT efforts in enabling women farmers for exposure visits, meetings with visitors, and explaining the success stories of SHGs in other parts of India enormously motivated them, as several of them underscored this point during our discussions.

By 2018, Kothapally had a total of 40 women SHGs, each with 10–12 members, thereby covering a total of 400 households out of the total 502 households in the village. To begin with, an SHG got a loan from a nearby nationalized bank (in this case corporation bank) of Rs 1 lakh in 2005–2006, which in turn was distributed to each member at Rs 10,000 to be repaid in 3 years in equal EMIs. By 2007–2008, the loan amount increased to Rs 2.5 lakh per SHG; thereby, each member got about Rs 25,000. By 2013, the loan amount was increased to Rs 5 lakh, thereby Rs 50,000 per member. Now, in a current scenario (year 2018), each member pays Rs 1000 to Rs 2000 per month as a repayment directly credited to their bank account. All members realized that improved repayment surely enhances their credit worthiness for their next round of credit. Hence, each member is keen to repay on time. In some cases, other members chip in to help in case a member defaults for a month or two. The SHG lends money to its members with an interest rate of 12% per year, and the same goes back to the commercial bank; thereby the SHG as such don't retain or create its own funds. Further, we found that during each loan round, women have used it largely for unproductive expenditure. But, from the household perspective, this was essential. For example Smt. Anusuya, whose SHG also distributes spent malt for all cattle owners, spent her first loan amount of Rs 10,000 on purchasing a color television in 2005. During the second round in 2007, Rs 25,000 was spent for agriculture implements and children's education. During the third round in 2013, Rs 50,000 was spent on purchasing gold ornaments. However her bank passbook showed that she is regularly repaying her monthly installments, and till March 2018, she has returned Rs 43,000. She is hopeful of repaying the remaining money (Rs 7000) in the next few months. Then her SHG is planning to jack up the loan size to Rs 7.5 lakhs. Thereby, each member gets Rs 75,000. In a big relaxing mood, she expressed that the ICRISAT interventions and its support have enabled all women (400 households) to come out of money lenders and high interest rates. They also feel grateful to the ICRISAT's efforts and at the same time feel very comfortable in gradually strengthening their family and lifestyle due to this financial support.

Before the watershed project in Kothapally, both women and men were not getting proper wages for agriculture work. But, after the watershed project was initiated, the male workers started getting wages of Rs 30 and female workers Rs 20 as daily wages. And for non-agriculture work, wages were Rs 50 for both. Working with the watershed project, they get to know their proper wages. Current (year 2018) daily wage rates are Rs 300 for men and Rs 250 for women (Fig. 11.8).

The SHGs have taken Rs 5 lakhs as a loan which is distributed per SHG. However, the loan amount varies depending upon the size of the group. The smaller groups (less than ten members) have taken Rs 2 lakhs as loan, while the big groups (with ten members or more) have taken Rs 5 lakhs as loan; thus a total of Rs two crores have been taken by 40 SHGs together, over the last 10 years.

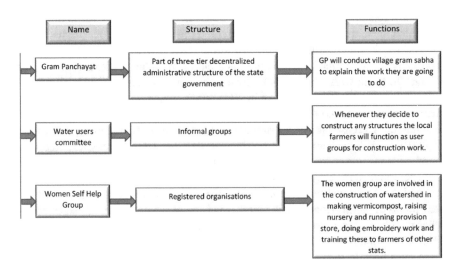

Fig. 11.8 Functions of self-help groups in Kothapally

Table 11.2 Spent malt lifting and sales in Kothapally village (from June 17, 2013)

Sl	Details	2013	2014	2015	2016	Jan–May 2017
1	Quantity of spent malt procured (kg)	186,180	431,222	533,004	628,400	325,466
2	Loss of quantity due to evaporation and other losses (kg)	23,620	21,326	35,460	38,240	22,119
3	Net quantity of spent malt sold during the year (kg)	162,560	409,896	497,544	590,196	303,347
4	Sale price of spent malt to farmers (Rs/kg)	3.5	3	3	3.17	4
5	Total income during the year (Rs)	535,100	1,229,688	1,492,632	1,834,458	123,388
6	Cost of spent malt paid to factory during the year (Rs)	325,815	754,639	932,757	1,167,860	895,031
7	Transport charges of spent malt from factory to village (Rs)	143,700	305,700	344,400	410,400	204,500
8	Equipment, labor, and other charges during the year(Rs)	19,450	115,000	81,000	80,000	39,000
9	Rent for storage and handling charges (Rs)	38,500	0	49,500	79,000	33,000
10	Total cost during the year (Rs)	527,465	1,175,339	1,407,657	1,737,260	1,171,531
11	Net benefit to the SHG (Rs)	7635	54,349.5	84,975	97,198	41,857

11.3.4 Spent Malt to Boost Milk Yield Levels

A women SHG at Tejasvi was promoted to buy the spent malt from nearby breweries such as M/S SABMiller. Three tons of spent malt is purchased every alternative day at the rate of Rs 2.75 per kg and sold at Rs 5 per kg. This selling price includes 20% loss incurred for the dryness of the malt from the stage of procurement to sales and the transportation cost from factory to village. The distance from Kothapally village to spent malt factory is approximately 40 kms (Table 11.2 and Fig. 11.9).

11.3.5 Land Structure

Kothapally has a total area of 464 ha (cultivable area 430 ha and irrigated area 200 ha) spread over in 242 survey numbers, and more than 20% of the land is cultivated by share croppers: Rs 4000–Rs 5000 per acre per year for dry land and Rs 7000–Rs 800 per acre per year for irrigated land. The share cropper has to pay money at the end of the crop season. There is another trend, leasing-out method with a period of 2–3 years. The villagers have not seen any contract farming. Majority of share croppers are small landholders.

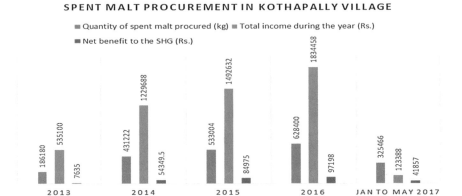

Fig. 11.9 Spent malt procurement in Kothapally village from 2013 to May 2017

11.4   Water Rights

Water rights and water allocation are gaining relevance due to increased competition among domestic, industrial, and agricultural water users. Formal laws and rules are increasingly being used to govern increasingly scarce water supplies, manage inequity in access (by individuals, communities, and riparian states), and distribute vital water resources (Raju and Sarma 2004). Nevertheless, questions of relative priorities among different uses remain: irrigation versus drinking water, rural versus urban demands, agricultural versus industrial demand, irrigation and power generation versus flood moderation, and abstractions for use versus maintenance of minimum flows. These are questions of socio-political-economic choices.

In the 1980s, India faced several river disputes, which have led to legal disputes between states, and these had caused law and order problems in several places. Commenting on this situation, Chattrapati Singh (1991) argues that the original natural rights over rivers and other natural waters belong to the people of India and not to the government or the state. He asserts that people have a natural or fundamental right over what is essential for their life which inherently belongs to them. Governments can have only a legal usufruct right, with the consent of the people. When the government acquires any usufruct right for specific public use, it should compensate the original users or beneficiaries and define the "public" as bearers of rights. He concludes that to make the state accountable and to make water use equitable for all in India, a number of amendments are required in the Easement Act, the Irrigation Laws, Panchayat and Municipal Corporation Laws, Water Supply Acts, and other laws related to water.

One source of difficulty is that India has taken over the colonial legal legacy in its entirety. The Constitution itself is largely based on the Government of India Act of 1935. In recent decades, attempts have been made to introduce elements, some traditional and some modern, that do not easily fit in with it or with one another. Water, as a basic right, is a useful idea, but it has the potential of being asserted not only against the state but also against the community or civil society. Iyer (2003) argues that one has to reconcile the individual fundamental right of all people to water as life support with the community's right of managing common poor resources. Both of these must be reconciled with the responsibilities (and therefore the rights) of the state to control, regulate, and legislate. Indeed, many interstate issues including the intra-basin apportionment and inter-basin transfer of water arise from different interpretations of the ownership rights in regard to water. Questions of rights relating to water or in the context of water resources arise in diverse ways and from different perspectives (Iyer 2003). Table 11.3 summarizes some of these perspectives.

The Government of Andhra Pradesh (United Andhra Pradesh) made the historic decision in January 1996 to transfer management of all irrigation systems to farmer's organizations. In 1996 and 1997, the government held several consultations with farmers of the major project areas, districts magistrates, media, universities, legislators, and parliamentarians, to evolve a strategy for the constitution of farmers' organizations in the irrigation sector. This series of consultations led to the enactment of the Andhra Pradesh Farmer's Management of Irrigation Systems Act of 1997 (APFMIS Act). After the bifurcation of United Andhra Pradesh, the Telangana state also, to a large extent, adopted this Act, with minor refinements. This applies to our study area of Kothapally village.

11.4.1 NGO Support

During 1999–2002 the watershed-related structures (check dams, farm ponds, sunken pits, water diversion canals) were constructed. These structures were designed and implemented by the government's Department of Agriculture under the watershed program. During the initial 4 months, the MV Foundation, an NGO, helped the local farmers to work on the watershed-related activities; after their withdrawal, the local Assistant Director of Agriculture has taken the help of a local organization, READ[2](Rural Education and Agricultural Development), an NGO. This NGO over the years, based on the walkthrough surveys and consultations with the local farmers, has been designing and executing this watershed project. It is on the same line under the scheme DPAP-IV (Drought-Prone Areas Programme) in Ranga Reddy district.

The READ organization was headed by Shri V. Naveen Kumar, a civil engineer who had previous experience in designing and executing the watershed structures; he has been working with the Department of Agriculture on 13 projects performing design and estimation of cost and supervising these executions. The organization

---

[2]The READ organization is a registered volunteer body. As a professional NGO, it adopts twin stages of demonstration models of development at the local level and also works for policy reforms at higher levels. Its main mission says "Establish sustainable regenerative rural lifestyles, environment protection and create replicable models per village."

Table 11.3 Questions of rights related to water in India

Type of perspective	What it means	Applicability and remarks	Examples
Riparian	Rights to waters of flowing river inherited or claimed by different users located alongside (or in the vicinity of) that river	At the level of households, farms, communities, villages, or towns, but occurs in a more marked form at the level of political or administrative units within a country, or at that of "co-riparian" countries	The Inter-state Cauvery Dispute, the Indus Treaty of 1960 (India-Pakistan), the Mahakali Treaty of 1996 (India-Nepal), and the Ganges Treaty of 1996 (India-Bangladesh)
Federalist	Distribution of rights and powers in relation to water between different levels in the federal structure	Three lists – union, state, and concurrent. A distribution of legislative power of the Union Parliament and State Legislature. 73rd and 74th amendments to the Constitution provide a three-tier structure	Villages, cities, and states. Water is listed as state subject in the Constitution
Formal law	Includes judicial determination. But quite complex and confusing	The right to drinking water is a fundamental right. Unclear whether legislative (and corresponding execution) powers, conferred by the Irrigation Acts, imply ownership of rivers and other surface waters by the state. Governments tend to assume so	In the Inter-state Water Disputes Act of 1956, "inter-state" means "inter-government." International treaties or agreements over rivers (e.g., the Indus Treaty, the Mahakali Treaty, the Ganges Treaty). Groundwater rights go with land ownership rights
Customary law	Communities allocate water according to land ownership or often investments made, caste, or community membership	Farmer-managed irrigation systems, domestic water supply systems not built by the government	Small tanks in southern India Kuhls in Himalayas, wells
Civil society	Arises in three different but inter-connected contexts where local communities are involved	(a) Efforts to protect people's rights, particularly poor, disadvantaged communities and tribal groups, from state and its agencies and large projects. (b) Move to revive traditional community-managed systems of water management. (c) New initiatives in social mobilization and transformation	Anna Hazare's in Ralegan Siddhi in Maharashatra or Tarun Bharat Singh's in Alwar District in Rajasthan

(continued)

Table 11.3 (continued)

Type of perspective	What it means	Applicability and remarks	Examples
Participatory	Various from full involvement of users from the early stages of planning to mere formality of asking for comments on a plan, program, or project prepared entirely within the government	Participation is invited in projects planned and implemented in a wholly non-participatory manner. Often the inability of the state to manage a project and provide the planned services. The state is usually unwilling to enter into a contractual relationship with users and accept binding obligations with penalties for non-performances	Large irrigation projects and tanks across the country. Some exceptions are WUAs in A.P. and recently some tanks in Karnataka

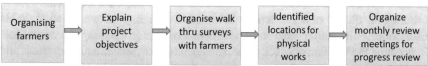

Fig. 11.10 Process followed by READ

was set up in 1997 and has offices in Hyderabad and Sangareddy. Based upon earlier experience, the READ follows the below shown process (Fig. 11.10).

This process was more of the inclusive nature and enabled regular interactions and taking care of suggestions and grievances of stakeholders. In addition to this the READ organization has taken care of identifying the exact locations for watershed structures with the help of ICRISAT scientists, getting approval from the government agencies, organizing farmers to execute the work with their labor, and effectively supervising the physical work on a day-to-day basis.

However, the whole process was not free from constraints of two types, social and technical. On the social side farmers had disagreements on the location of structures of check dams and percolation pond construction of canal for diverting water to open wells; several meetings were organized frequently to handle these situations. Interestingly 80% of the stakeholders used to participate in these meetings which enabled reaching consensus. On the technical side, the NGO faces the problem of undulated terrain, depth of soil, and hard rock zone. Based upon the scientific suggestions given by the ICRISAT scientist, the NGO could be able to resolve the structural designs and execute them.

During the construction of check dams and other structures, the concerned stakeholder groups contributed labor and were paid on the basis of actual physical work, which became a good source of income for local farmers. This had also enabled a saving of Rs three lakhs in total expenditure, allocated for all watershed structures. The saved money has been retained even after 14 years. The interest earned from this saved money is used for essential expenditure related to some repair and maintenance of these structures.

Generally, below each check dam, up to half a kilometer length, both on left and right side, open well owners and the farm land owners became members of the watershed user committee. They also became primary beneficiaries of the stored water through recharge of their groundwater. In recent years the READ and ICRISAT scientists are thinking about preparing a water budget taking into consideration all watershed structures put together in the village.

11.4.2 Watershed User Committee

As part of constructing various check dams, in Kothapally, the ICRISAT in consultation with the gram panchayat and local villagers had set up several watershed user groups: one for each check dam. These user groups played a significant role in ensuring the quality of construction of check dams; sharing labor ensured equitable distribution of irrigation water to all members. The user groups played a critical role in deciding the location of check dams, sunken pits, farm ponds, and farm bunds and creation and usage of vermicompost and usage of appropriate inputs for crops.

As a part of this study, we interviewed a few members of the watershed user committee, to understand the perceptions, the execution process, and the benefits realized over the years. Shri Narayan Reddy, 70 years old, has been a member of these user groups from the beginning. He is still active in farming practices and his fields are spread over in different locations totaling up to 70 acres. This includes the land purchased in recent years by his two sons. In his humble submission, he admitted that, in 1999 when the state government allocated some funds to promote watershed activity in this village, "We did not know what to do and how to do it. A local NGO (READ) helped to understand it somewhat better but more critical was the help received from ICRISAT, who explained towards the step-by-step process both orally and through maps and pictures. The ICRISAT helped us in setting up the overall watershed committee users groups, training them, collecting member fee enabling the members to actively participate at all stages of execution and reaping its benefits. The funds created during those years are still with the farmers and we feel proud of it. Because of the transparency in the entire mechanism and no wasteful expenditure was made in the entire process."

The initial positive impact had changed the mindset of local farmers. The first check dam construction in 1999 and its retention of water in the upstream was the first positive impression. Then recharging of 14 dried-up open wells in the downstream and retention of water in these open wells even during the summer had deeply caught farmers' attention. Particularly, the technical assistance provided by ICRISAT had boosted up farmers' confidence in ICRISAT's works. Shri Narayan

Reddy said that they just followed ICRISAT's suggestions at every stage and started reaping benefits in manifold. All the key interventions like structures, crop varieties, packages, and practices have led to enormous enhancements in their crop yield levels, more fodder for their cattle, and higher milk yield levels. All these naturally resulted to earning higher income by all households; he further added that "even landless agricultural labours could be able to get more jobs and higher wages; in turn our own active participation, both in construction and other related civil works, enabled us to save in costs out of the total grant given of Rs 16 lakhs; we spent only Rs 12 lakhs and we got more work got done with this limited money. When we estimated in a normal contactor mode, the total cost was more than Rs 30 lakhs; now we see ourselves the high cost incurred in check dam structures and in other areas constructed by contractor mode; yet those structures are constructed by quality problems and damages; whenever a group leader has shown indiscipline or not maintained transparency in his dealings, the micro-level committee has changed the particular person in the interest of overall group."

Mr. Narsimha Reddy, 70 years old, was the first president of GP serving from 1990 to 2003; later he worked as president of the Adarsh Watershed Association, Kothapally. He explained how they started the Kothapally watershed; initially the Andhra Pradesh government sanctioned Rs 20 lakhs, out of which the government had deducted Rs 4 lakhs. Of the remaining Rs 16 lakhs, Rs 14 lakhs was spent for watershed activities, and Rs 2 lakhs was given to BAIF for artificial insemination activity. They collected Rs 1 lakh from the village farmers as development fund and deposited it in a commercial bank for maintenance of structures in the future. The watershed association was established in 1999, with a five-member team (Mr. Narsimha Reddy, president; Mr. Azam, member; Mr. M. Sathaiah, member; Mr. Parmaiah, member; and Mrs. Vara Laxmi, member).

11.4.3 APFMIS Act

The APFMIS Act provides for the establishment of water user associations in the irrigation sector. Projects have been classified as minor (less than 2000 ha), medium (2000–10,000 ha), and major (more than 10,000 ha). Rules pertaining to the water rights of member and farmers' organizations are mentioned in Table 11.4.

After reviewing the performance of the farmers' organizations during the last 5 years, they decided to change the WUA organization and to add certain amendments to the Act. Public debate was initiated on the changes in the WUA setup to carry out necessary revisions to the APMIS Act. The amendments required to the Act were brought through Act 7 of 2003 in April 2003. Later in 2015, according to the Act and rules, the objectives of the farmers' organizations shall be to promote and secure distribution of water among its users, to adequately maintain the irriga-

Table 11.4 Rights of WUA members and WUAs

Particulars	Rights of WUA members	Rights of WUAs
Quantity of water	As per specified quota	Receive in bulk from WRD
Tradable water rights	Can transfer to any user within the WUA operational area	
Water deliveries	Suggest improvements	
Information on	Water availability, allocations, canal operation timings and duration	Water availability, opening/ closing of main canal, periods of supply and quantity
Crop choice	Full freedom but within water allocated	
Allocation of water		To both members and non-members
Distribution of water		Among water users on agreed terms of equity and social justice
Layout of field channels and drains		Suggest improvements/modifications to enable all farmers to have access to water
Groundwater		Plan and promote groundwater use

Source: Based on APFMIS Act 1997 and rules, as issued by the Government of Andhra Pradesh

tion system, to utilize water in an efficient and economical manner, to encourage the modernization of agriculture, to optimize agricultural production, to protect the environment, and to ensure ecological balance by involving the farmers, instilling a sense of ownership of the irrigation systems in accordance with the water budget and the operation plan.

The APFMIS Act of 1997 and several rules issued by the state government in recent years have led to the establishment of different kinds of rights. These are broadly riparian rights and traditional or customary rights. The clearest water rights are water use rights. These may not be absolute rights in terms of quantities, but they are general rights to draw water within the Warabandi schedule to make the schedule applicable to the command of WUA. Warabandi means a system of distribution of water to users by turn, according to an approved schedule indicating the day, the duration, and the time of supply. This stems from the fact that the right to water basically depends on the availability of water in the source and this is subject to vagaries of nature.

There is a provision in Panchayat Raj bodies of Telangana, Andhra Pradesh, and Karnataka states that vested fishery rights in minor irrigation tanks to gram panchayats (village council). There is a conflict of legal rights between WUAs and gram panchayats. The government sought to solve this conflict by entrusting the auctioning of fishery rights to the fishery department and distributing the proceeds to WUAs and gram panchayats as per the ratio prescribed.

11.4.4 Water Rights Changes and Conflicts

11.4.5 Tank Resources

Another example of conflict about rights to water resources and their management occurs in the case of tanks. The revenue sources from the tank include irrigation fee, leasing-out fee for sand mining, and leasing-out fee for tank bed cultivation. The revenue generated from the use of tanks goes to different agencies. The legal framework and state policies in the four southern states (Telangana, Andhra Pradesh, Karnataka, and Tamil Nadu) are not clear. Owing to this, both users and the agencies get into conflicts. Table 11.5 shows some of the legal conflicts in the case of Karnataka. Similar issues also apply in Tamil Nadu, Telangana, and Andhra Pradesh. Due to these conflicts, tank user associations (both formal and informal) are facing constraints on mobilizing resources, threatening the survival of tank user associations (TUAs).

The conflict between irrigation and fish culture needs is increasing. Inflows into the tank have decreased due to overexploitation of groundwater and expansion of agriculture in catchment areas as well as construction of new water-harvesting structures in catchment areas. Owing to conflicting rights to the resources, tank user associations are in a dilemma in several places. This occurs even after the states have come up with a clear policy to support tank user associations and transfer management of tanks to the user groups. These states did not focus adequately on the legal implications. For example, in Telangana and Andhra Pradesh, some TUAs have been drawn into court cases by local village councils based on claims made on fishing rights. This is also true in the emerging TUAs in Karnataka. While Telangana and Andhra Pradesh have issued a government order making provision for TUA

Table 11.5 Groundwater rights in Telangana, Andhra Pradesh, and Karnataka

Groundwater rights	Legal frameworks introduced	What it provides	Relevant states
Groundwater use	Water, Land and Trees Act of 2001	Registration of existing wells and permission for new wells	Telangana and Andhra Pradesh
	(Government of Andhra Pradesh)	State can close down existing wells if they are found to be causing damage to environment	
	Groundwater Bill of 2002 (Government of Karnataka)	State can permanently close down tube wells used for agriculture purposes if they are affecting drinking water wells in rural areas	Karnataka

Source: (a) Government of Andhra Pradesh 2004 and (b) Kolavalli and Raju 2003

Table 11.6 Conflicting legal frameworks over tank resources

Use and source of income	Agency responsible and focus of the conflict
1 Water fee	Imposed by irrigation department and collected by revenue department
2 Fishing	Fisheries department auctions fishing rights. Generally, a trader sub-leases it at a much higher amount to a fishing group. No preference to TUA
3 Silt (soils rich in nutrients are taken by local farmers)	Mines and geology department has control and ownership
4 Tree nurseries and plantations in the tank bed and catchment area	Forest department claims rights
5 Ownership and management of all water bodies in the Constitution, gram panchayats	According to the 73rd amendment of the Indian Constitution on the village revenue boundary, they have rights

Source: Based on APFMIS Act of 1997 and various rules, as issued by the Government of Andhra Pradesh

rights over fishing, in practice, it could not happen. Karnataka is now drafting a comprehensive Act on tanks in southern India (Table 11.6).

11.4.6 Organizations for Revitalization of Water Bodies

Extensive research in member organization in various fields suggests that robust self-governing user organizations achieve high levels of goal-cohesiveness, governance effectiveness, and member-need responsiveness by satisfying the following four design principles. These principles are linked to user organizations for tank revitalization in Rajasthan by Raju and Shah (2000). We found, in Kothapally village, that the watershed user committee is following these principles to a large extent.

(a) **Member-centrality of the goal:** Members give their allegiance and loyalty to a user organization only to the extent that it serves their interest. If the purpose is non-central or non-immediate for members, the organization generally proves stillborn unless externally propped up. If its purpose is non-central to members, people do participate but withhold their support as soon as selective inducements in the form of subsidies, wage labor, and other giveaways are withdrawn.

(b) **Goal-cohesive governance:** Most boards/management committees of induced user organizations tend to be powerless and divisive and pursue agenda that have vague or no relationship to members' stake in the organization. Often boards/management committees do not genuinely represent member interest or are dominated by outside agencies. It also happens that, once elected/nominated, members have no or ineffective mechanisms to hold a non-performing board accountable. Training and education for effective board development can help promote member-centered governance, but the design features that are really critical include (a) election of all or most members of the governance structure by members, (b) stake-based voting rights and representation, (c) members' rights of recalling non-performing boards/members, and (d) board/management committee (and not the catalyst organization) as the custodian of the decision-making authority of the general body.

(c) **Get the rights operating system:** The operating system is the vehicle through which the user organization serves members. It has to be so designed that it can generate high rewards for collective action. Operating systems, which successfully add value, often invent new methods of organizing resource system or getting tasks performed or service delivered compared to the existing methods. The operating system for a successful user organization creates unique member allegiance propositions that provide members strong and continuing reasons to offer their loyalty and allegiance to the organization and comply with the established behavioral norms.

(d) **Secure and retain member faith and allegiance:** The user organization fails when the members desert it. On the contrary, a user organization becomes increasingly stronger as it amasses the allegiance and loyalty of a growing membership. Even when well designed, a user organization has to be launched carefully since adverse member expectations formed from early experience take a long time to undo. The operating system thus has to create positive member expectations at the start and meet them so that members develop faith in its capacity to deliver. Virtuous cycles of this type strengthen member allegiance and faith in the user organization which is the sure formula for its success and centrality.

11.4.7   Legal Issues Related to Water Sharing

It is essential to deal with critical legal issues for law related to water distribution and water rights. These are fundamental to irrigation water law and will need to be resolved; their resolutions are likely to affect irrigation management transfer (Brewer et al. 1999).

(a) **Irrigators' Water Rights**

At present, most states in India have enacted laws giving the state government the right to allocate surface water to uses and users as it judges what is best to be adapted for public interest. For surface water, the legal situation is that the state can decide what the rights of an irrigator are. Every state attracts surface irrigation water rights to specified pieces of land, although the rights specified are often quite different. No such laws exist for groundwater. Implicitly, then, groundwater is legally controlled by the holder of the land overlying the ground-water. Since water rights are tied to land, water rights can be transferred from one individual to another only through the transfer of land.

Is this the best situation? What should be the rights of an irrigator to water for irrigation? Here we discuss three key issues.

- Limitations to governments' power over surface water – Present laws give the government almost complete power to decide who should have access to surface water for irrigation. However, a few court cases have attempted to limit this power (Singh 1991). Though the government's right to regulate irrigation is paramount and sovereign in character, it should not be exercised arbitrarily. First, the government's rights should be subject to the irrigator getting the same quantity which he or she is accustomed to. Second, any changes should be based on clearly enunciated and publicly accepted principles, such as equity of distribution, or priority for drinking water. Security of water rights is essential to give confidence to irrigators so that they will invest in irrigated agriculture. Also, when changes in water rights occur, they should be seen as principled and follow well-established rules rather than arbitrary and subject to private or personal influence.
- Limitations to landholders' power over groundwater – Given rapidly falling water tables in some areas and documented inequalities in availability of groundwater, for public interest, there is a need to provide some legal limits to the power of landholders to exploit groundwater. Even Kothapally farmers, during our walkthrough survey, expressed similar concerns.
- Tradable water rights – Recently, some specialists have begun advocating the development of tradable water rights as a means of ensuring that water is used most effectively. Tradable water rights are rights to use water that can be transferred all or in part, separately from the transfer of land. While tradable water rights themselves should be permanent or very long term, to ensure their security, the transfer of water rights need not be permanent: water rights can be leased for a season, a year, or many years. Establishing markets in tradable water rights may offer many benefits, including empowerment of water users, provision of investment incentives, improved water efficiency, reduced incentives to degrade the environments, acceptability to farmers, improved equity in the provision and financing of water services, and increased flexibility in resources allocation. Assigning tradable water rights to individuals within WUAs or to the WUAs themselves enhances the control of these groups over water resources, ensuring better access to water.

- The present situation offers little security of water rights for either surface water or groundwater. In the case of surface water, there are no clear principles or rights to balance government power. In the case of groundwater, since landholders have an unrestricted right to pump, no one has assurance that groundwater will be available in the future. Some clear specifications of irrigators' and the government's' water rights would be helpful.

(b) **Irrigation and *Panchayati Raj***

As part of the process of decentralization of power, the 73rd amendment to the Constitution (Government of India 1993) was passed in 1993 to include provision for *Panchayati Raj*. Rural local governments have now been vested with independent power and resources. Most importantly, it is the intention of the amendment to give local *panchayats* the power to control all natural resources within their jurisdiction, including water.

This leads to a key question of the powers of local *panchayats* over irrigation. In the case of large-scale government-managed systems covering the jurisdictions of more than one local panchayat, it is likely that the state will reserve basic management powers for the state irrigation agencies. However, this amendment could potentially lead to panchayats being given powers to control the functioning of WUAs within the village or block, particularly those that manage small irrigation systems such as tank systems. It may also lead to panchayats taxing irrigation. The draft irrigation bill in Kerala, for example, allows a local authority to levy water cess with the previous sanction of government. In Kothapally, during the discussions with the president of gram panchayat, less awareness was indicated, among president and GP members.

The relationship of local panchayats' control over water resources and the powers of other management bodies, particularly WUAs, may need legal explication.

Other helpful features that could be included in law or in regulations include:

- The law or government regulation should encourage and support catalytic agents for limited periods to bring farmers together initially and help them organize themselves and get official recognition.
- The law or regulations should define ways to recognize WUAs without putting undue burdens on them in registering and meeting state requirements.
- The law or regulations should require that the by-laws of WUAs include means by which members can change the leadership at reasonable intervals or, when needed, based on their opinions.

Responsibilities and rights of WUAs that might be specified in regulations or in an agreement between each WUA and the state government should include:

- WUA responsibilities

  (a) Equitable distribution of water among its member and non-members.
  (b) Operate, maintain, and repair the system within the WUA area.
  (c) Pay water fees to the state.

- WUA rights

    (a) Right to get an agreed quantity of water and information on delivery schedules.
    (b) Right to manage their own finances, including the rights to charge irrigators water fee, to set their own fee levels, and to mobilize additional resources through other means.
    (c) Right to punish/fine defaulters, preferably including the right to stop delivery of water to defaulters.
    (d) Right to resolve conflicts within WUA areas.

These changes are clearly not enough to bring about irrigation management transfers. Legal amendments are necessary but not sufficient. In practice, the effects of legal acts may be small. Indeed, most of the progress has been made without proper legal support. Legal support, however, is essential for the long-run viability of irrigation management transfer.

11.4.8 Check Dams and Groundwater Recharge

Over 4 years' time (2000–2004), a total of 14 check dams were constructed across two different water streams within the Kothapally village boundary. These check dams have a distance of 400–500 m, between two check dams, while the seven are on the one stream and the rest on the other stream. The village has a total of 88 bore wells and 62 open wells. In recent years, the water storage in the check dams has enabled the chart of all the 62 open wells recharged with 30–40 ft depth; some 50 bore wells with a depth of 150–200 ft also get recharged on a regular basis. Interestingly all the 88 bore wells in the village are drilled after the construction of check dams.

As part of water resources regulation stored in the check dams, the watershed association has formally decided, as recorded in their proceedings, that the stored water will not be lifted without permission, either by machine or by manual implements. These decisions are also approved by the *gram sabha*. Should there be any violation of these decisions, it attracts a penalty of Rs 25,000.However, as regards the extent of pumping out bore wells or open wells, located on individual farm plots, there is no restriction on quantity of water being lifted and crop pattern.

In recent years under MNREGA (Mahatma Gandhi National Rural Employment Guarantee Act), labor is used for desilting the water storage area in the check dams and repairing the apron. This desiltation is carried out once in 2 years in nine check dams. All farmers have utilized the desilted soil on their own for farm lands. The decantation has also helped to store more water and for a longer time. Before decantation the water used to dry up by the end of December or early January; after decantation water remains in the check dams till the end of March. This has boosted up the recharge levels, particularly in the open wells.

The electricity availability over the first 12 months has increased to round the clock; earlier it was constrained by 8–12 h per day. A low tariff[3] (at Rs 30 per irrigation pond, respective of the horsepower size) is implemented. Indeed, farmers are not expecting this low-rate tariff. Owing to this low tariff plus 24-h availability, majority of farmers have installed auto-starter; thereby, majority of the pumps run round the clock leading to over-irrigation and wastage of precious water.

11.5 Policies for Strengthening Institutions

(a) Government policies to enable local village councils and suitable local institutions to regulate and promote ecosystem services with more clarity on role of institutions, individuals, and property rights. (b) Encourage more and deeper studies in understanding the need for appropriate institutions (both formal and informal) for sustainable ecosystem services and evolve guidelines for better decision-making process. (c) Flexibility for region-specific programs/activities for local ecology-based sustainable provision of ecosystem services.

This study of Kothapally village in the semi-arid tropics of southern India has scope to develop methodology involved in the collection of primary data on ecosystem services in the region, identifying institutions involved in ecosystem maintenance of the region, environmental benefits enjoyed by the different stakeholders, and interdependence of localities, institutions, and ecosystem services of the region. By this there is potential to draw attention toward where the system is going wrong and where it can be corrected through economic sense, environmental perspective, and sociological way of institutions working on it, for proper maintenance and sustainable use and conservation of resources at local/micro-level. These micro-level studies possess environmental benefits at global level revealing that the work of local-level institutions is very important for achieving these benefits and more interest should be given to improve these local-level institutions.

---

[3] In other states of India, tariff rates range from Rs 100 per HP per month to a flat tariff of Rs 900 for 10 HP per month. The Telangana government owing to this low tariff is also incurring Rs 3200 crores per year (?) for supplying power to the farmers.

Kothapally Milk Production During 2017–18

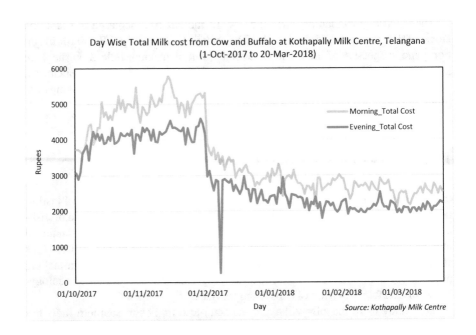

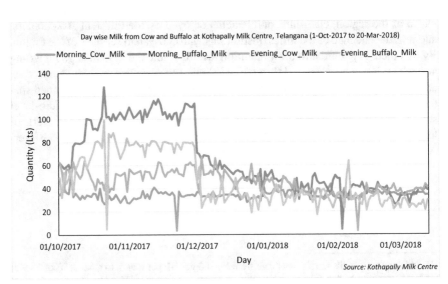

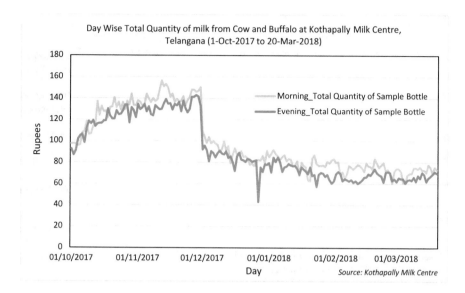

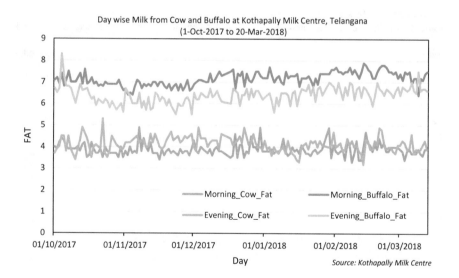

References

Adarsha watershed, Kothapally, India – An innovative and upscalable approach. In *Research towards integrated natural resources management: Examples of research problems, approaches and partnerships in action in the CGIAR*, 2003.

Brewer, J.Kolavalli, S., Kalro, A.H., Naik, G., Ramnarayan, S., Raju, K.V., & Sakthivadivel, R. (1999). *Irrigation management transfer in India. Policies, processes and performance*. Oxford: IBH Publishing Co. Pvt Ltd.

Bromley, D. W. (1991). *Environment and economy: Property rights and public policy*. Oxford: Basel Blackwell.

Government of Telangana. (2018). Stree Nidhi Credit Co-operative Federation Ltd., Annual Report 2016–17.

Hanna, S. S. (1998). Institutions for marine ecosystem: Economic incentive and fishery management. *Ecological Application, 8*(1) Supplement pg, S170–S174, Ecological Society of America.

Iyer, R. R. (2003). *Water; perspective, issues, concerns*. New Delhi: Sage.

Kolavalli, S., & Raju, K. V. (2003). *Protecting drinking water sources: A sub-basin view* (Working Paper No. 126). Bangalore: Institute for Social and Economic Change.

North, D. C. (1992). *Transaction costs, institutions, and economic performance*. San Francisco: Institution for Contemporary Studies.

Raju, K. V., & Sarma, C. V. S. K. (2004). *Water rights in India and water sector reforms in Andhra Pradesh* (Working Paper 149). Institute for Social and Economic Change.

Raju, K. V., & Shah, T. (2000, June 3). Revitalisation of irrigation tanks in Rajasthan. *Economic and Political Weekly*.

Raju, K. V., Wani, S., & Poornima, S. (2014). Institutions for ecosystems services in semi-arid tropics: A study of a cluster of villages in South India.

Singh, C. (1991). *Water rights and principles of water resources management*. New Delhi: India Law Institute.

Chapter 12
Summary and Way Forward

S. P. Wani and K. V. Raju

Abstract The vast semi-arid tropics (SAT) area covering 120 million ha in Asia is also the home for 852 million poor and 644 million food- and nutrition-insecure people. Growing water scarcity and increasing land degradation in the dryland SAT areas are further aggravated due to impacts of climate change. In order to transform the dryland areas, innovative integrated watershed management model was developed and piloted by the International Crops Research Institute for the Semi-Arid Tropics (ICRISAT) in partnership through consortium approach, convergence with the government programs, collective action and cooperation (4Cs) approach. How resilience of the communities was built through integrated watershed approach encompassing the livelihoods is described fully. The outlines of different chapters indicate briefly the strategy, and various aspects including the process adopted and its impacts are covered.

Keywords Climate change · Resilience · Holistic solutions · Integrated watersheds · Partnerships and consortium · Livelihoods · Communities

Promoting Research for Development The ICRISAT was working in watershed development since 1972 with Vertisol Technology and piloted on farmers' fields in different agro-eco regions. However, it was not scaled-up/adopted by the farmers in spite of involvement of concerned state government agencies. In 1995, a multidisciplinary team of scientists' assessment of watershed studies in different agro-eco region pilot/benchmark sites was undertaken. Low adoption of Vertisol technology although demonstrated on farmers' fields was due to poor participation of the farmers as the approach adopted was contractual participation, and one-size-fits-all approach was adopted. The new multidisciplinary experiment on station in Vertic

S. P. Wani (✉)
Former Director, Research Program Asia and ICRISAT Development Centre, International Crops Research Institute for the Semi-Arid Tropics (ICRISAT), Hyderabad, Telangana, India

K. V. Raju
Former Theme Leader, Policy and Impact, Research Program-Asia, International Crops Research Institute for the Semi-Arid Tropics (ICRISAT), Hyderabad, Telangana, India

© Springer Nature Switzerland AG 2020
S. P. Wani, K. V. Raju (eds.), *Community and Climate Resilience in the Semi-Arid Tropics*, https://doi.org/10.1007/978-3-030-29918-7_12

Inceptisols demonstrated that using integrated watershed management approach, these soils can be cropped during two seasons. As desired and demanded by the district officials, Kothapally watershed was selected based on severe water scarcity, large extent of rainfed areas and community's need and willingness to participate in the program through full ownership/participation. The journey of innovation in Kothapally and how it became an exemplary (Adarsha) watershed with different strategies adopted is described. It evolved by the consortium of research institutions, government department, non-government organization and the farmers' community. The drivers of success are identified, and complete journey of innovation through a detailed timeline is covered in the chapter.

Cope Up with Climate Change Knowledge of climate and weather helps in devising suitable strategies and managing crops to take advantage of the favourable weather conditions and minimizing risks due to adverse weather conditions. Role of climate assumes greater importance in the semi-arid rainfed regions where moisture regime during the cropping season is strongly dependent on the quantum and distribution of rainfall vis-à-vis the soil water holding capacity and water release characteristics. Evidences over the past few decades show that significant changes in climate are taking place all over the world as a result of enhanced human activities through deforestation, emission of various greenhouse gases (GHGs) and indiscriminate use of fossil fuels. Various studies show that climate change in India is real, and it is one of the major challenges faced by Indian Agriculture. Agro-climatic analyses of the watersheds based on long-term weather data include concepts of rainfall probability, dry and wet spells, water balance, length of growing period (LGP), occurrence of droughts, climate variability and projected climate change. Long-term weather data of Kothapally watershed was obtained from installed automatic weather station and Indian Meteorological Department (IMD) gridded data and analysed for characterizing the agro-climate and assessing the climate variability. Results indicated clear increasing trends in temperature and considerable changes in rainfall. Climate projections also indicated large changes in temperature and rainfall at Kothapally in the future. Implementing Integrated Watershed Management Programme in a holistic way can mitigate the adverse effects of climate variability and change and enhance the capacity of small-farm holders to manage extremes of drought and floods in a sustainable way.

Integrated Soil Health Management Kothapally watershed is a typical representative of rainfed (800 mm rainfall) semi-arid tropics (SAT) with varying soil depth in the watershed, widespread soil degradation as the major challenge coupled with low crop yields and family incomes. Before the onset of initiative during 1999, soil health mapping and baseline surveys showed varying soil depth in fields at different toposequence, macro-/micronutrient deficiencies along with low soil carbon (C) levels and heavy soil loss through erosion that compromised with crop production in the watershed. Inappropriate fertilizer management decisions leading to negative budget for primary nutrients in major crops/cropping-systems highlighted suboptimal fertilizers use. Unawareness about micro-/secondary nutrient deficiencies like sulphur (S), boron (B) and zinc (Zn) and lack of addition of such fertilizers contributed to low crop yields and declining fertilizer and water use efficiency. Farmers' participatory trials highlighted yield loss of 13–39% in crops like sorghum and maize in the absence of deficient micro-/secondary nutrient fertilizers. Recycling of on-farm wastes through vermicomposting and biomass generation using N-rich *Gliricidia* on farm boundaries were promoted for fertilizer savings and crop yield benefit alongside soil carbon building for developing resilience. The impact of integrated soil health management practices cumulatively observed over 13 years was demonstrated during 2012 soil health mapping that showed improved mean level of soil organic C, available nutrients, viz. phosphorus (P), B, Zn, S and significantly reduced number of fields with low nutrient/C levels. Along with yield advantage, soil loss was significantly reduced from 3.48 t ha^{-1} in untreated area to 1.62 t ha^{-1} in treated watershed area.

Agricultural Water Management (AWM)   interventions in Kothapally watershed enhanced provisional, regulating and supporting ecosystem services. Kothapally watershed was in degraded stage before 1999 and is transformed into highly productive stage through science-led natural resource management interventions. A number of AWM interventions, such as field bunding, low-cost gully control structures, earthen check dams, masonry check dams, etc., were built as per hydrological assessment and need of the community. Ridge to valley approach of rainwater harvesting addressed equity issue as farmers from upstream end benefited along with downstream users. A number of AWM interventions reduced surface runoff (30–60%) and soil loss (two- to five fold) and enhanced groundwater recharge (50–150%) and base flow. Water table increased from 2.5 to 6.0 m on an average after the AWM interventions. This change has translated into surplus irrigation water availability and crop intensification especially during post monsoonal season. Further, all such changes translated into better crop yield, higher cropping intensity, higher crop production and net income over the years that resulted into building the resilience of the individuals and community to cope with droughts and impacts of climate change. This case study clearly indicates that large untapped potential exists in dryland areas which could be harnessed through science-led NRM interventions. Scaling-up approach of these interventions through pilots at various locations in India, Thailand, Vietnam and China demonstrated the potential for overcoming food and water scarcity sustainably at the same time contributing to meet the Sustainable Development Goals of zero hunger, water availability and climate actions.

Improved Livelihoods   Climate change presents an additional challenge for sustainable food production in the developing world. It is necessary to enhance present yield levels. Deriving and popularising suitable cropping systems is critical. Integrated watershed management (IWM) implemented in Adarsha watershed, Kothapally, is a fine example of overcoming negative impacts. About 250 rainwater harvesting structures were constructed. About 27% of the rainfall contributed to groundwater recharge and risen the groundwater levels by 2.5–6 m. Due to increased water availability, farmers were able to diversify crops and grow two/three crops. In the post-rainy season, rice, sorghum and chickpea were grown. Increased soil organic carbon (SOC) stocks were observed due to inclusion of legumes in cropping systems. According to the Intergovernmental Panel on Climate Change (IPCC), climate change affects crop production more negatively than positively, and countries like India are highly vulnerable. As per the General Circulation Model (GCM) CESM1-CAM5, RCP8.5, for Kothapally area, maximum and minimum temperatures during monsoon (Jun–Sep) are to be increased by 1.0 and 0.9 °C, respectively, by 2030. At the same time, rainfall in June and July together expected to decrease by 65 mm, and August and September together are projected to have increased rainfall of about 50 mm. Sustainability of crops like cotton, maize, sorghum and pigeon pea was studied using crop simulation models. Impacts of climate change on productivity were estimated and suitable adaptation strategies were derived.

Achieving Positive Impacts Adarsha watershed is a successful scientific narrative of sustainable integrated watershed programme conceptualised by ICRISAT for efficient management of natural resources. Creating a proof-of-concept and a learning site for extension agents, NGOs, the national agricultural research system, policy makers as well as for farmers was one of the main objectives of ICRISAT when the institute started its work in the Adarsha watershed in Kothapally village, Ranga Reddy district, in Telangana, India, in 1999. Water harvesting structures, 14 check dams, 97 gully control structures of loose stones, 1 gabion structure and others together have created a net storage capacity of 21,000 m^3 which harvested nearly 70,000 m^3 runoff water per year and have brought an additional area of 55 ha into irrigation by improving the groundwater table from 2.5 to 6.0 m. With improved technologies, farmers obtained high maize yields (28%) than the base year. Cotton has observed major yield gain (387%), major because of both technological change (Bt cotton) and assured water availability. Pigeon pea has recorded an increased productivity over the timeline (61%). Watershed has contributed to improved resilience of agricultural income despite the high incidence of drought during 2002 in the watershed. Whilst drought-induced shocks reduced the average share of crop income in the non-watershed area from 44% to 12%, this share remained unchanged at about 36% in the watershed area. Livestock sector also contributed significantly to total household income in watershed villages even during drought situations. Reduction in marginal cost due to supply shift has improved the cost-benefit ratio across the crops and ranged from 1.72 in cotton to 4.1 in pigeon pea. The BCR is worked out to be more than 2 and IRR 31%, implying that the returns to public investment such as watershed development activities were feasible and economically remunerative. The NPV worked out to be Rs. 141 lakh INR for the entire watershed. The total treated area in the watershed was around 465 ha, and the NPV per ha worked out to be Rs. 30,000 INR implied that the benefits from watershed development were higher than the cost of investment of the watershed development programs. The study revealed that the watershed development has the potential for poverty reduction by generating impressive returns on investment even during drought year. The new generation watershed intervention emphasises achieving the food and income security of farmers whilst maintaining the integrity of the eco-hydrology and other natural systems in the watershed.

Tracking Land Use Changes Is Critical Land use analysis, cropping pattern and sources of irrigation are important for planning and improving the rural economic aspects. Geospatial method was adopted for deriving and analysing the land use at micro-level in Kothapally village covering parts of Adarsha watershed. The village is characterised by large number of marginal holdings, and the average holding of 0.96 ha is lesser than that in Telangana state. Kothapally is characterised by agriculture and the land put to non-agricultural use is 9.31%. Migration to urban area, selling road side land for commercial purpose and alienated land not being cultivated by poor have been found to be main reasons for fallow lands. Cropping systems comprises of cereals, pulses, fruits, commercial crops, vegetables, flowers and plantations. Over the years, extent of cereals area has remained almost same with sorghum cultivation almost stopped. Pigeon pea appears to be gaining prominence and area under cotton has doubled. Sugarcane has disappeared. Rainfall is the only source of irrigation in Kothapally. Groundwater is being used for irrigating 109.49 ha. Groundwater withdrawal is on the increase and shallow open wells are becoming unusable. Different types of rainwater harvesting have been adopted. Though interventions have yielded better results, more demand for water has caused more exploitation of groundwater. The village has all the required basic infrastructural facilities, and the wastewater treatment facilities built as part of watershed management interventions are appreciated by the community. Almost all the houses have sanitary facilities and are connected with underground drainage.

Enabling Women Involvement at All Stages Is Pivotal Despite the fact that women are the world's principal food producers and providers, they have long been deprived of their due share and identity. Kothapally is one of the initial watershed projects that demonstrated on ground that a holistic development model not only conserves natural resources for sustainable productivity and income improvement but also harnesses the synergies to tailor the benefits in mainstreaming women farmers. This has showcased the model to focus on selective activities that directly benefit women. Some important activities that increase incomes of women revolve around interventions like milk production, kitchen gardens, composting, value addition, non-farm livelihoods through capacity building, collectivization and market linkages.

Increase Income Levels Enormous efforts made over the years have enabled increased family income by 28% over the non-watershed villages due to integrated watershed development livelihood model. Kothapally is a unique peri-urban village in the vicinity of Hyderabad with 400 households mainly cultivators. Further, it has also demonstrated that through watershed development resilience of the communities during drought years was built and no migration took place as share of crop income remained constant, whereas in non-watershed villages it dropped by 75%, and people even the farming households had to migrate for their livelihoods, and proportion of income from non-agriculture activities increased dramatically to 74%. Suitability of integrated watershed livelihood approach strongly indicated the approach to be scaled-up not only for increasing production and income but also for contributing substantially to the SDGs, viz. zero hunger, reducing poverty, climate change interventions and women empowerment.

Strengthen Institutions and Governance Mechanism The critical analysis of type of institutions existing and functioning over the years in this *gram panchayat* (village council) has shown how they are governed as organizations, both formal and informal. The governance mechanism evolved over the years, and its refinements are made over time; their structures for governance are analysed. The literature related to Kothapally, available till recently, has rarely focused on rural institutions and their governance; over the last 17 years (2000–2017), several publications, based on both desk review and field survey, were published. Its review indicates scant attention on institutions and governance mechanism. Thereby, authors have emphasised on this dimension, based on extensive discussions with stakeholders and walkthrough survey in farmer fields. In the course of mapping the rural organizations and institutions in Kothapally, the authors have tracked: (a) *gram panchayat*, (b) women self-help groups (SHGs), (c) watershed user committees, (d) watershed association, (e) farmers' association, (f) local NGO, (g) water rights and water distribution mechanism. The authors held a series of discussions with various stakeholders, both on one-to-one and in a group, spread over several weeks in Kothapally village. Additionally, they made a content analysis of the records maintained by the *gram panchayat*, SHGs and watershed user committees. Authors have critically analysed the institutional mechanism, management process, key organizations and their role, including the pivotal role played by the local village council, how local women groups through their self-help groups successfully dealt with micro-financing and enabled convergence with other programmes. Also described are utilization of spent malt to boost up milk production, how water rights were tactfully managed with support from a local nongovernmental organization and methods followed by watershed users' committee for revitalization of water bodies.

12.1   Way Forward

This account of transforming water scarce area of Kothapally into prosperous agricultural area on the periphery of Hyderabad where farmers have resisted the temptation to sell their land for non-agricultural use is an eye-opener as well as very encouraging case study to scale up this model into large areas. This journey of innovation was evolutionary and boldly adopted the changes constantly to enlarge the approach to meet the needs with changing times and priorities and ensuring sustainable development for building resilience of the communities for the climate change. This innovation encompassed the empowerment of community through demand-driven science-based interventions for conserving and efficiently using the scarce natural resources and transformed the village through adoption of livelihood approach since 1999 in the watershed management. Based on exemplary development of the Kothapally, through consortium approach the convergence with government schemes is achieved and has enabled the community to work collectively to harness the power of togetherness ensuring sustainability through capacity building.

Promote Innovative Methods As envisaged before initiating the watershed development, the scientists and the community made this watershed famous across the country and spread its success in Africa and other parts of rainfed areas in the world. In India this case study played a major role in the development of New Watershed Guidelines embracing the elements of livelihood approach against the traditional soil and water conservation structure-driven approach; participatory planning and monitoring; use of new science tools like GIS; remote sensing (RS); simulation modelling; involvement of scientists in nodal committees at national, state and district levels to guide planning and monitoring; and most importantly inclusion of microenterprises for women and landless people in the rural areas. Increased water availability enhanced crop yields as well as community diversified their cropping systems and livelihood systems along with crop intensification dynamically considering the increased family incomes. So far so good, but this journey of innovation has to continue as the changing times are posing new and unexperienced challenges like climate change, urbanisation, dwindling water resource, decreasing family farm sizes and changing food habits, feminization of agriculture, and most importantly with education of the new generation the farming has to transform to meet their expectations and also ensure food and nutrition security for the ever-growing population in the country. As this journey of innovation was dynamic, evolutionary and so became exemplary, watershed is and will be looked on for guidance as well as for solutions in future.

Let us take some examples for way forward to address the emerging challenges. The best part is this model has been scaled up in different agro-ecoregions with varying socioeconomic swath, and several actors are active in developing the solutions for emerging challenges. The important emerging challenge is of climate change which brings in the changes in agro-ecologies as well as new pests and diseases forcing the communities to change their cropping systems as well as food systems. This model has already shown how community reacted to climate change, as in the year 2002 this village received 200 mm rainfall in a day and not any water harvesting structure in the village recorded any damage. Similarly, in the year 2000, the village successfully faced the drought situation and since 1999 the villagers have never faced shortage of drinking water, whereas till 1999 the women had to fetch water from January onwards from long distance as all the open wells in the village were dried. When *Helicoverpa* increased the cost of cultivation of cotton, farmers shifted quickly to maize pigeon pea intercrop system which was more remunerative than cotton. As soon as Bt cotton came in the market, farmers shifted quickly to cotton and almost 99% cotton grown in the village is Bt cotton. The community also demonstrated their thirst for seeking the information about the climate change from the scientists in order to adapt to the changes. This clearly suggests that in order to build the resilience of the communities, we need to empower them to take their decisions based on the scientific and validated information. Whilst scaling up or adopting to this approach at any location, it will be important that communities are supported by the scientific institutions to provide updated evidence-based information about the climate change so that they can do the needful adaptation/changes in their practices.

Support Digital Technologies In order to empower the communities through scientific knowledge and new communication technologies such as information technology (IT), Internet of things (IoT), artificial intelligence (AI), machine learning (ML) and use of big data for developing decision support systems (DSSs), we need to develop platforms to disseminate the advisories using mobile phones as 95% of rural households in India have mobile phones. There is an urgent need to develop integrated IT platforms to develop scientific advisories to the farmers holistically as against the current approach of several players pushing the half-cooked advisories for specific interventions. As we are aware that increased productivity and production alone does not guarantee increased incomes for the farmers, they need right information and linkages to market their produce to realise better prices. As currently intermediaries corner 60–65% share in the price consumer pays, there is an urgent need to adopt value chain approach to get better share for the farmers in the price paid by the consumers. This only can be achieved when scale of operation can be achieved, and generation after generation the farm sizes are becoming smaller and smaller.

Promote Farmer Producer Organizations The scale of operation can be achieved through collectivisation by adopting farmer producer organisations (FPOs) which the government of India is promoting. However, to make the FPOs functional and sustainable, needed changes in the policies are essential to be brought in by various states as agriculture is a state subject. Good examples from sustained FPOs need to be studied, and as Government of Andhra Pradesh brought out the new FPO Policy and started adopting it, we need state-specific enabling policies to make the FPOs functional and sustainable. Once good FPOs are there, these can help the farmers to collectivise the planned production of specific commodities as per the ecological potential and needed by the market as well as establish the linkages with input suppliers and product purchasing companies to increase their family incomes. The sustainable FPOs can also adopt the value chain approach and generate employment in rural areas by promoting small and medium-sized enterprises (SMEs). We need to make small farm holder as entrepreneur, and farming needs to be transformed into a business model for making agriculture viable livelihood and preferred option as against the current no choice situation for the farmers.

Encourage Youth and Agribusiness As the educated youths will like to get into agriculture as a business proposition, the agriculture will need to become smart agriculture/precision agriculture where most operations can be handled through smart gadgets as well as mechanisation. Mechanisation can be only feasible for small farm holders through efficient and cost-effective machine hiring centres run as a business model by the entrepreneurs with supporting maintenance and delivery systems. Efficiency in all operations will need to be enhanced dramatically as water and land resources are becoming scarce day by day, and further due to climate change, unusual events of heavy rains, long and frequent dry spells and extreme low and high temperatures stress, there is a need for urgent and effective interventions for enhancing water and nutrient efficiencies. Precise and efficient use of nutrient and water applications is critical for handling the climate change impacts in long term.

Rural microenterprises must be developed as agriculture alone cannot satisfy the needs of farming communities in India and other developing countries in Asia and Africa. In future, agriculture will have to become smart and mechanised and be run as a business model which will be a part-time occupation for the educated youths. In order to enable them to undertake farming, the rural microenterprises will need to be developed, and enabling policies to support such interventions to run farms as a business model is needed. In brief, the future agriculture will have to be dynamic, smart and holistic and be run as a business model by the empowered communities to achieve the food and nutrition security along with improved farm incomes for the ever-growing population to achieve sustainable development.

Printed in the United States
by Baker & Taylor Publisher Services